Where We Live, Work and Play

Praeger Series in
Transformational Politics and Political Science

The Politics of Transformation: Local Activism in the Peace and Environmental Movements
Betty H. Zisk

The Latino Family and the Politics of Transformation
David T. Abalos

Mediation, Citizen Empowerment, and Transformational Politics
Edward W. Schwerin

Strategies of Transformation Toward a Multicultural Society: Fulfilling the Story of Democracy
David T. Abalos

Beyond Confrontation: Transforming the New World Order
Charles Hauss

Teaching Democracy by Being Democratic
Theodore L. Becker and Richard A. Couto, editors

Transforming Privacy: A Transpersonal Philosophy of Rights
Stefano Scoglio

Where We Live, Work and Play

The Environmental Justice Movement and the Struggle for a New Environmentalism

Patrick Novotny

Praeger Series in Transformational Politics and Political Science
Theodore L. Becker, Series Adviser

PRAEGER

Westport, Connecticut
London

Library of Congress Cataloging-in-Publication Data

Novotny, Patrick.
Where we live, work and play : the environmental justice movement and the struggle for a new environmentalism / by Patrick Novotny.
p. cm.—(Praeger series in transformational politics and political science, ISSN 1061–5261)
Includes bibliographical references and index.
ISBN 0–275–96026–9 (alk. paper)
1. Environmental justice—United States. 2. Environmentalism—United States. I. Title. II. Series.
GE180.N68 2000
363.7′03—dc21 99–086115

British Library Cataloguing in Publication Data is available.

Library of Congress Catalog Card Number: 99–086115
ISBN: 0–275–96026–9
ISSN: 1061–5261

First published in 2000

Praeger Publishers, 88 Post Road West, Westport, CT 06881
An imprint of Greenwood Publishing Group, Inc.
www.praeger.com

Printed in the United States of America

The paper used in this book complies with the Permanent Paper Standard issued by the National Information Standards Organization (Z39.48–1984).

10 9 8 7 6 5 4 3 2 1

Copyright Acknowledgment

The author and publisher gratefully acknowledge permission for use of the following material:

"Demonstrator's Song," reprinted from *The Indianapolis Star*, September 1982.

We've got a road to walk, it's mighty steep too.
But one thing that we know is true,
Let history not be repeated,
A people united will never be defeated.

Our good old earth we've got to guard and share,
We've got to keep her safe and free from care,
And that means standing up for what is right.
We'll fight the poison with all our might.

We won't stop, oh Lord, we'll barely rest.
We're committed 'cause the truth is our test.
We have righteousness on our side,
Those poison devils had better hide.

Some folks think we'll never win,
That toxic waste is a deadly sin,
But we ain't gonna let nobody turn us around.
We once were lost, but now we're found.

Demonstrator's Song
Warren County, North Carolina
September 1982

To my parents
John and Margaret Novotny
and to my grandmothers
Dora Kleinert and Catherine Novotny

Contents

Acknowledgments

I would like to thank the many people who have given me the intellectual and the personal encouragement to complete this book. Too many people have given too much of their time, energy, and effort to thank here in mere words alone. I would, however, like to thank those family members, friends, and colleagues who have helped see this book come to fruition.

My parents, John and Margaret, have been with me every step of the way in a book that really began many years ago. They instilled in me an intellectual curiosity that is with me to this day. From the days of my childhood when they encouraged me to put aside my books and enjoy the outdoors, to our summer trips to Estes Park, Colorado, my parents have always guided me with their love and their encouragement to ask the big questions in life. Their careful guidance and encouragement in supporting me emotionally, financially, and intellectually in the pursuit of my studies has been, without a doubt, the greatest gift and the truest blessing in my lifetime.

To the loving memory of my grandmothers, Dora Kleinert and Catherine Novotny, this work is also dedicated. Their encouragement, their pride in their grandson, and their love have guided me every step of the way and been a constant blessing in my life. To all of my uncles, aunts, and cousins and to my entire family, I want to thank you all for the many special memories and the love which you have shown me over the years. The world is held together by the bonds of family, and I have been truly blessed with a family that has never wavered in its love and concern for one another.

My colleagues at Georgia Southern University have been an inspiration in their teaching, their scholarship, and their friendship. My affiliation these past years with the University would not have been as meaningful to me were it not enriched by the friendship of Dr. Lane Van Tassell. The friendship and the laughter shared with Dr. Bob Dick has also been an especially meaningful part of my experience there. Bob has been the consummate freethinking scholar—always willing to question the Zeitgeist—and most of all, a true friend. I also thank Drs. Debra Sabia, Lois Duke Whitaker, Roger Pajari, Sudha Ratan, Charles Gossett, John Daily, Shafik Hashmi, George Cox, Rebecca Davis, Karen McCurdy, and Chris Ludowise, all of whom are especially supportive and irreplaceable colleagues in the Department of Political Science at Georgia Southern University. Drs. Steven Weiss, William McIntosh, Darin Van Tassell, and other colleagues at the University have been kind friends and supportive throughout. I have been fortunate to have had the encouragement in my teaching and scholarly pursuits of two outstanding Presidents as well, Dr. Nicholas Henry and Dr. Bruce Grube, both of whom have done their utmost to nurture scholarship here at the University. My good friend Dr. George Shriver and the rest of the 10:00 A.M. coffee group also merit a heartfelt thanks for their warm collegiality and friendship over the years. Georgia Southern University is a wonderfully diverse environment, and I thank each and every one of my colleagues there for their encouragement and their support.

The University of Wisconsin and the city of Madison, Wisconsin were a wonderful starting point for the intellectual journey of this book. To the Department of Political Science at the University of Wisconsin, I want to give my heartfelt thanks. I want to especially thank Dr. Virginia Sapiro—for your scholarship, your advice, and your professional encouragement when I needed it most, I will forever owe you a debt of gratitude. I want to also thank Drs. Benjamin Marquez, Graham Wilson, Paul Lichterman, and Harvey Jacobs for their careful and thoughtful guidance of my work. My former professors and colleagues in the Department of Political Science in North Hall are too numerous to mention here, but one name in particular stands out for encouraging a creative and imaginative approach to the study of politics and culture. I want to thank Dr. Murray Edelman whose spark of imagination and intellectual curiosity has been with me since my first day as a student in his Empirical Political Theory Seminar.

The University of Wisconsin left a profound impact on both my intellectual and scholarly work and on my own life as well. From the unforgettable late spring afternoons on the shores of Lake Mendota to the hours studying in Memorial Library and in the reading room of the Wisconsin State Historical Society, my scholarship was nurtured in Madison. Friendships and graduate school colleagues are far too numerous to mention here but the friendship of Heather Hartley was a constant source of enjoyment, laugher, and inspiration in my years in Madison. I want to also thank Allen Hunter and my friends at the A.E. Havens Center for the Study of Social Structure and Social Change at the University of Wisconsin.

I would also like to recognize those associated with Greenwood Publishing Group. Dr. Theodore Becker of the Department of Political Science at Auburn University served as the catalyst for this book with Greenwood and I hope that he will not be disappointed with this contribution to his outstanding series. Heather Staines has provided invaluable encouragement and support at every stage of production. And no acknowledgment would be complete without my special thanks to Deborah Whitford, whose meticulous and careful work on the book has made it all possible. Deborah has patiently worked with me throughout production and it is her steady hand in the final stages of this project that has truly allowed this book to come together.

My final thanks are to Theresa Beebe, my loving friend, my partner, and my confidant who has been with me since the beginning of my writing this book. Theresa has always provided a sympathetic ear and has offered her love and support at every step. I have enjoyed our weekend trips exploring the back roads of Georgia and the streets of Savannah with all of its mystery and charm. Our afternoon and evening walks under the Spanish moss and live oaks of the historic squares of Savannah left this book richer with a love that grew in that time and that continues to grow each day.

Introduction

> Environmental justice is a fundamental human right.
>
> —The Reverend Jesse Jackson

In the past decade, poor and working-class communities, particularly African-American, Native American, Latino, and Asian Pacific communities, have formed the environmental justice movement. This movement is configuring a new understanding of the environment that connects it with many other concerns. The movement does not separate environmental hazards, affordable housing, health care, and racial discrimination from each other. Instead, what it terms "environmental racism" is understood as yet another form of discrimination that results in a disproportionate incidence of environmental problems in African-American and Latino communities. These problems are caused by what are seen as discriminatory land use practices; decisions by industry, which locate hazardous wastes in these communities; and the uneven enforcement of environmental regulations by federal, state, and local officials; all of which, in turn, are described as part of a wider history of discrimination.

The environmental justice movement is responsible for the emerging awareness of environmental problems as they affect the poor, particularly African-American, Latino, Native American, and Asian Pacific communities. Study after study confirms that the poor bear the burden not only of hazardous wastes, but of excess carbon dioxide and ozone, and increased cases of asthma and lead poisoning. "Being poor in America," explained Greenpeace in a December 1991 report, "means breathing foul air, working filthy jobs, and living next to toxic waste landfills and incinerators."

The connections between the environment and impoverishment are evident to millions of people who live in the vicinity of waste sites, described by Greenpeace as poverty's silent partners.

The enormity of the threats to the environment has been thoroughly documented over the past three decades, from Rachel Carson's pathbreaking studies in the early 1960s of the health effects of pesticides to more recent concerns about the effects of pollutants on the human immune system. That the poor and persons of color suffer even more from these effects, writes Luke Cole of the California Rural Legal Assistance Foundation, "is confirmed in local and national studies of the impact of toxic production and disposal, garbage dumps, air pollution, lead poisoning, pesticides, occupational hazards, noise pollution, and rat bites" (Cole 1992, 623–624). Even with this evidence, it took the events in a rural North Carolina county in the late 1970s to focus the attention of civil rights lawyers and advocacy groups on the way that these communities suffer from environmental problems.

In December 1978, state officials made the decision to site a landfill in the sparsely populated rural community of Warren County, North Carolina. More than thirty-one thousand gallons of waste had been dumped along roadsides by an electronic transformer firm in Raleigh in what constituted the largest illegal dumping of polychlorinated biphenyls, or PCBs, in the nation's history. The state of North Carolina announced its decision to bury these contaminants at a landfill in rural Warren County. As an unincorporated rural area with an African-American population of nearly seventy-five percent and the lowest per capita income in the state, Warren County had few regulations to ensure that the landfill would be held accountable to local residents.

The residents of Warren County fought the construction of the landfill for nearly four years. At public gatherings, they expressed their concern about the health effects of the site, particularly because many residents drew their water from wells that could be easily tainted by groundwater contamination. Within days of the first shipment of materials to the landfill, residents who had opposed the landfill through public hearings, petitions, and letter writing campaigns turned to a campaign of civil disobedience reminiscent of the civil rights movement.

The protests in this rural North Carolina community resonated with the history of the civil rights movement in the South, as reflected in Jenny LaBalme's photographic essay *A Road to Walk: A Struggle for Environmental Justice*. Protests centered around the Coley Springs Baptist Church, whose congregation led protests during almost two months of demonstrations and civil disobedience. "The fervor of the days," writes LaBalme, "hung heavy that September and October. Dull yellow sunlight pressed through the frosted windows of the Coley Springs Baptist Church. People's legs stuck to the wooden church pews and their arms flapped back and forth holding fans splashed with watercolor portraits of Martin Luther King, Jr.

and John and Bobby Kennedy. Parents, grandparents, and schoolchildren squeezed into the rows of pews and turned their faces toward the podium the way they always did before marching" (LaBalme 1987, 9). Arrests occurred every day for six weeks while trucks brought in the contaminated soil to the landfill. Men, women, the elderly, and children lay on the ground in front of these trucks while North Carolina state police arrested demonstrators and carried them away from the site.

On the dust-filled roads leading to the landfill in Warren County, civil rights leaders protested alongside local residents. More than five hundred demonstrators were arrested, including Reverend Benjamin F. Chavis, Jr. of the United Church of Christ Commission for Racial Justice, Reverend Joseph Lowery of the Southern Christian Leadership Conference, Floyd McKissick of the Congress for Racial Equality, and the District of Columbia's Walter Fauntroy. Lois Marie Gibbs, the well-known organizer of the Love Canal Homeowners Association and founder of the Citizen's Clearinghouse for Hazardous Wastes, joined the residents of Warren County, bringing the concerns of race and poverty raised by the demonstrations in North Carolina to the hazardous waste movement.

The participation of national civil rights leaders not only bolstered the confidence of local residents, but made the protests in Warren County a definitive moment in the emergence of the environmental justice movement. In the end, the protests were unable to prevent the siting of the landfill. Through a coalition of groups, however, the demonstrators did help bring together the community, which elected the county's first African-American sheriff and later prevented the state of North Carolina from making Warren County a permanent waste site. The demonstrations also brought the landfill to the attention of members of the Congressional Black Caucus, leading to a study by the U.S. General Accounting Office of the socioeconomic and racial composition of communities with hazardous waste landfills. The study found that three of the four largest landfills in the United States were in predominantly African-American communities (U. S. General Accounting Office, 1983). It also found that twenty- five percent or more of the population in communities with at least one hazardous waste landfill were below the poverty level.

The protests in Warren County drew the attention of civil rights leaders and generated additional studies of the relationship between race and poverty, and the incidence of environmental problems. Perhaps the most important legacy of the protest in North Carolina is that it prompted the United Church of Christ Commission for Racial Justice to conduct its own study of hazardous wastes. Reverend Benjamin F. Chavis, Jr., executive director of the United Church of Christ Commission for Racial Justice, termed this "environmental racism," rallying groups around the principle that "there can be no clean environment without a just environment."

The impact of hazardous waste sites, landfills, and other facilities in the context of impoverishment and substandard housing is a crucible of problems for communities that already lack adequate health insurance or medical facilities. An estimated forty million people live in the vicinity of hazardous waste sites in the United States. Of these, many are thought to be uninsured or without access to affordable health care (Lewis, Keating and Russell 1992). With little or no access to adequate health care, the consequences of contaminated water and air are even greater for these uninsured people living in the vicinity of environmental hazards.

The theoretical perspective of framing is postulated in this book as necessary for understanding the politicization of the environment in the environmental justice movement. Framing is the way that the leaders in a movement assign meaning to and interpret problems in such a way as to mobilize participants. Framing is a part of what is described by some scholars as a dynamic process by which a movement and its struggle for social change is connected with a larger set of cultural values, beliefs, and practices. What distinguishes framing from other parts of a political movement is that it encompasses the culture and even the language that is used in a movement.

The Labor/Community Strategy Center in Los Angeles, Local 4-620 of the Oil, Chemical and Atomic Workers in Geismar, Louisiana, the Gulf Coast Tenants Organization in New Orleans, and the SouthWest Organizing Project in Albuquerque are each studied in this book as groups that frame the environment in ways that make it a part of wider struggles in their communities. The environment is framed as intermingling with many other community issues. Each of these groups has formed its own understanding of the environment in ways that have much in common with the histories of the civil rights, farmworkers, and labor movements. The perspective of framing, it is argued in the subsequent chapters of this book, is necessary for understanding the process whereby the Gulf Coast Tenants Organization, the Labor/Community Strategy Center, the Oil, Chemical and Atomic Workers Local 4-620, and the SouthWest Organizing Project have become involved with the environment of their communities. Each of these groups has transformed the language they used to express their understanding of the environment in order to connect it more closely with their previous work. This effort to define the environment in a language more closely associated with the historical struggles of these groups is not merely semantic. It has an immediate connection with the organizing of these groups.

1

Where We Live, Work and Play: Framing, Political Mobilization and Environmentalism in the Environmental Justice Movement

The public awareness of the environmental hazards of pesticides, automobile emissions, and the contamination of the nation's rivers and lakes—the most famous of which was the eutrophication of Lake Erie in 1972—was part of the changing public sentiments in the decades following World War II. Widespread fears of health threats from the atmospheric testing of nuclear weapons increased concern for the environment as early as the 1940s. However, the emergence of the postwar environmental movement did not get fully under way until later in the next decade, when the magnitude of environmental depredation was given renewed attention in the work of scientists, naturalists, and journalists, who publicized the magnitude of environmental damage during this period through their writings and photographs.

The environmental movement that emerged in the early 1960s and the growth in membership of groups such as the Sierra Club reflected the growth of the postwar middle-class as much as the changing attitudes toward the environment in the wake of World War II. With the expansion of state and national parks, the growing enthusiasm for recreation and outdoor activities, and the popularity of widely read works such as Rachel Carson's *Silent Spring*, millions of middle-class Americans widened the constituency of the environmental movement. Whereas the conservationist and preservationist movements in the early twentieth century were for the most part comprised of scientific figures, wealthy philanthropists, and government leaders, the postwar environmental movement arose from a much broader segment of society.

The establishment of the Environmental Defense Fund, the Environmental Action Foundation, and other new environmental organizations as well as the regulatory and legislative initiatives of the Water Resources

Planning Act of 1965, the Clean Air Act of 1970, and the Endangered Species Conservation Act of 1973 reflected a growing awareness of the environment. Environmental disasters such as the immolation of the Cuyahoga River in Ohio and the oil spill in Santa Barbara, California in January 1969 led to the ground swell of public opinion that culminated in the passage of the nation's most comprehensive environmental legislation to that date, the National Environmental Policy Act, and the formation of the U.S. Environmental Protection Agency as well as the establishment of a host of state agencies and departments concerned with environmental quality.

The postwar environmental movement and the protection of the nation's water and air demanded as a result were the mainstays of a movement unmistakably middle-class in its composition. The movement became associated with a college-educated, suburban middle class that was too often removed from the realities of the impoverishment of millions of Americans. The environmental movement also emerged during the height of the civil rights movement, when the attention of most organizations in poor and predominantly African-American and Latino communities was focused on other concerns.[1] While Rachel Carson's *Silent Spring* was a national best-seller, activists in these communities worked to draw attention to the racial strife and impoverishment that persisted as the underside of postwar economic prosperity.

For many years, the literature on environmental activism has purportedly found that lower income communities are less concerned with the environment. In this literature, environmental activism is rarely considered relevant to the lives of the poor, for whom the seemingly insurmountable problems of impoverishment and unemployment are presumed to outweigh whatever environmental concerns exist in their communities. There is also widespread agreement that higher socioeconomic status is strongly related to environmental activism. Research for the most part tends to describe the poor as indifferent toward the environment.

The emergence of a movement that is focused on the environmental problems of poor, working-class, and predominantly African-American, Asian Pacific, and Latino communities is doing much to challenge the assumption that these communities are indifferent toward the environment. The environmental justice movement is characterized by the leadership of tenants' associations, civil rights groups, and labor unions with little prior involvement in environmental issues. With years of experience in political organizing and well-established histories of working with racial discrimination and poverty, the leaders of these groups are redefining the environment as a way of encompassing many of the problems in their communities. This redefinition of the environment in poor, working-class and African-American, Asian Pacific, Native American, and Latino communities means that affordable housing, community services, and other problems not typically thought of as a part of the environment become incorporated into a new definition of the environment as "where we live, work and play."

Few studies of this growing environmental activism in predominantly African-American and Latino communities appreciate the extent to which the leaders of these groups are fashioning a new understanding of the environment that incorporates the concerns of communities not otherwise associated with and, in many instances, deeply mistrustful of the postwar environmental movement. This book is a study of the way in which the leaders of these groups are defining the environment as "where we live, work and play," incorporating it into their work for affordable housing, occupational health, and workplace safety. It is argued in subsequent chapters of this book that their framing of the environment is an indispensable part of the involvement of these communities with environmental problems.

POST-MATERIALISM, ENVIRONMENTAL CONCERN AND THE POSTWAR ENVIRONMENTAL MOVEMENT

While public opinion research has acknowledged a widening concern with the environment among many different groups, relatively little is known about the concern for the environment in poor communities and the way that these communities are involved with this issue. For decades, much of the sociological and political science literature has understood environmentalism as a part of the socioeconomic change from an industrial economy to a postindustrial economy. The literature widely assumes that the poor are less likely to be concerned with the environment than the college-educated middle class.

For years, scholars have understood environmentalism as part of a silent revolution in industrial societies (Inglehart 1977). Still others have seen the postwar environmental movement as a "totally new political cleavage" or as a "vanguard for a new society" (Milbrath 1984; Lowe and Rudig 1986). What all of these theoretical perspectives share is an assumption that a societal change in cultural values is a necessary part of environmental activism and that these values are more often than not grounded in socioeconomic transformations away from a reliance on manufacturing industry toward a post-industrial economy. It also purports to show an indifference to the environment among the poor, working people, and people of color (Hershey and Hill 1977, 439–458).[2] "In spite of a general lack of evidence," write Bunyan Bryant and Paul Mohai, "the conventional wisdom has been that people of color are not concerned about environmental quality issues" (Bryant and Mohai 1992, 2–3). That environmental organizations have, indeed, tended to have fewer members from poor, working-class, and African-American, Native American, and Latino communities is clearly borne out in nearly every study of the postwar environmental movement.

The widespread assumption that poor and predominantly African-American and Latino communities are uninterested in the environment has been challenged in the last decade, as an emerging body of public opinion literature has shown environmental concern to be more widespread than previously thought (Mohai 1990, 744–765). While this literature has found environmental concern to be widespread, there nonetheless remains the assumption that the poor and the working-class are not as involved with the environment as other groups in society.[3] That environmental activists tend to be drawn from the middle-class is a well-established finding in the literature on the postwar environmental movement, yet this too often draws away attention from those instances where less privileged communities are connecting their concern with the environment with other issues.

The theoretical perspective of framing is useful for understanding the extent to which working people and the poor have found new ways of connecting the environment with many other concerns in their communities, in part by organizing through labor unions, tenants' associations, civil rights groups, neighborhood associations, and other groups. This involvement of African-American and Latino communities with the environment has as much to do with the way that it is framed in these communities as it does with the availability of socioeconomic resources or political opportunities alone. It is argued in this book that when the environment is connected with longstanding grievances in these communities as well as cast in such immediately understandable terms as racism, rights, or injustice, involvement with the environment is more readily engaged by groups in these communities.

FRAMING AND THE STUDY OF POLITICAL MOVEMENTS

The emergence of political movements is a perennial area of interest in the study of political science, sociology, and history. Although there is a well-established scholarly literature in the study of political movements, there are clearly any number of factors that have not been taken into consideration in the study of these movements over the years, in particular the processes involved in bringing a problem to the attention of particular groups as well as the language used by the leaders of a movement to depict the groups' problems.

Much of the research in the study of political movements is shaped by the resource mobilization perspective, which finds that groups are involved in movements because of the availability of material or institutional resources. Resource mobilization theory holds that the success or failure of a movement hinges on the ability of its leaders to draw in external resources or to make decisions that maximize the influence of a movement. A more recent approach includes political factors, which are less prominent in resource mobilization theory. This approach widens the scope of resources to include the political system itself and the opportunities that arise for groups to challenge it (McAdam 1982). While this approach is useful for understanding the circumstances that contribute to the emergence of a movement, it is also necessary to understand the way in which the leaders of a movement redefine or otherwise alter language commonly used in order to describe a particular problem as well as to connect it with their history and their previous organizing.

Much of the literature in the study of political movements is focused on the factors that either facilitate or impede the formation of organizations. Much less work, however, looks at mobilization that draws from groups that already exist.[4] The study of such organizations, in turn, raises important questions regarding the way in which the leaders of these organizations form their understandings of new problems as they arise as well as how they depict these problems in a way that makes them more closely associated with the organizing of these groups.

The involvement of tenants' associations, civil rights groups, and labor unions in the environmental justice movement and their incorporation of the environment through its framing as a problem of racism, injustice, or inequality is postulated throughout this book as indispensable to the movement. Through the definition of these problems as environmental racism, environmental justice, or even as issues of municipal infrastructure or workplace safety, it is argued that the environmental justice movement has been able to involve groups that would otherwise not be likely to work with the environment or think of themselves as environmental groups.

The organizations that are the focus of this book include the SouthWest Organizing Project in Albuquerque, the Gulf Coast Tenants Organization in New Orleans, the Labor/Community Strategy Center in Los Angeles, and the Oil, Chemical and Atomic Workers Local 4-620 in Geismar, Louisiana. These groups are widely acknowledged by activists and academic researchers alike as among the leading organizations in the environmental justice movement. Prior to their involvement with the environment, each of these groups worked for years on organizing in their respective communities.

The Gulf Coast Tenants Organization and the SouthWest Organizing Project, along with the Oil, Chemical and Atomic Workers and the Labor/Community Strategy Center, had established histories of organizing their communities around other issues prior to their work with the environment. In each case, these groups have sought to frame the environmental issues as part of the problems in their communities. From careful readings of their literature and publications, it is possible to understand the dynamics whereby the leaders of these groups have expanded their work to include the environment.

The Gulf Coast Tenants Organization has worked with African-Americans in the public housing projects and towns along the Mississippi River, many of which have been devastated by poverty, environmental hazards, and health problems. The SouthWest Organizing Project, which has been involved in many efforts to confront the environmental problems in the Southwest, began its work with the residents of Albuquerque's low-income and working-class neighborhoods. The Gulf Coast Tenants Organization and the SouthWest Organizing Project are involved in a redefinition of the environment around not only the environmental problems of their communities but the cultural identity of these communities as well.

The environmental justice movement has also included working-class communities and labor unions. Amidst the expanse of Louisiana's wetlands along the Mississippi River, members of the Oil, Chemical and Atomic Workers Local 4-620 and their families led a campaign of workers, community residents, and environmental groups against the longest lockout of a labor union in the nation's history. The work of Local 4-620 to incorporate the environment in Louisiana reflects the union's effort to frame the environment in these communities as part of more immediately recognizable concerns of residents, particularly an inequitable property tax system and the lack of an adequate community infrastructure in the towns along the Mississippi River. The involvement of the Labor/Community Strategy Center with the environment in Los Angeles is another instance of this kind. Many of its organizers emerged from the five-year campaign of the United Auto Workers to keep open a General Motors automobile plant in Van Nuys.

The Gulf Coast Tenants Organization, the SouthWest Organizing Project, the Oil, Chemical and Atomic Workers Local 4-620, and the Labor/Community Strategy Center are studied in the subsequent chapters of this book to ascertain their connection of the environment with their workplaces and community organizing. Although each of these groups has been involved with the environment in varying contexts and settings, with problems ranging from groundwater contamination and industrial accidents to workplace hazards and public transportation, there is nonetheless a similarity among the ways in which these groups are framing their understanding of the environment to connect it with racial discrimination, economic impoverishment, and political disempowerment in their communities.

THE MOVEMENT AS PROLOGUE: PROTEST TRADITIONS AND POLITICAL MOBILIZATION

Understanding the historical embeddedness of movements is a necessary part of understanding the framing of political movements. Protest traditions, or what have been referred to elsewhere as traditions of struggle, cultures of resistance, or repertoires of contention, are an important part of the framing in movements (Tarrow 1994; Snow and Oliver 1995; McAdam 1994; Morris 1984).[5] A movement often draws from the most salient historical experiences of a particular group.[5] These historical experiences are an especially important way of framing in a political movement.

The historical linkages of movements provide insight into the formation of a protest tradition as well as the means by which framing connects a problem with the consciousness of a particular group. It is argued here that movement leaders draw from the experiences of previous movements in their framing. "To be effective," writes Robert Benford, "a movement's claims not only need to resonate with the experiences of its audiences, they must also correspond with the cultural narrations, stories, myths, and folk tales of their culture" (Benford 1993b, 693). A protest tradition generates an even deeper sense of connection among participants by embedding a movement and its struggles in history. The music, leaflets, slogans, newsletters, and other literature of a movement facilitate among activists a connection with previous movements and reinforce their sense of identification and solidarity.

The history of civil rights, labor, tenants, farmworkers, and related movements is a significant part of the environmental justice movement. The SouthWest Organizing Project, the Labor/Community Strategy Center, the Gulf Coast Tenants Organization, and the Oil, Chemical and Atomic Workers Local 4-620 have each used the history of the civil rights, farmworkers, labor, and other movements as a part of the framing of their struggles with environmental issues. The leaders of these groups have gone to considerable lengths to give their organizing a continuity with that of preceding movements, in particular, the civil rights movement as well as the struggles of the United Farm Workers and the United Auto Workers. It will be argued in subsequent chapters that these groups are more likely to trace their involvement with the environment to the history of the civil rights, farmworkers, and labor movements rather than to the postwar environmental movement.

THE POLITICAL MOVEMENT AS DRAMATURGY, OR, THE PERFORMATIVE ELEMENTS OF POLITICAL ACTIVISM

The environmental justice movement's definition of the environment as "where we live, work and play" is postulated here as reflective of the importance of the framing in political movements. Movement leaders are not only understood as couriers of "extant ideas and meanings" but as actively engaged in the formation of new descriptive terms for movement participants (Snow and Benford 1992, 133–155; Benford and Hunt 1992, 36–55; Hunt, Benford and Snow 1994, 185–208; Mooney and Hunt 1996, 177–197; McAdam 1994, 36–57). Movement leaders use framing to challenge existing understandings of problems by bringing different interpretations, definitions, explanations, or understandings of these problems to the fore.[6] Framing is, in a very real sense, a part of the repertoire of mobilization strategies that are available to a movement, so that the movement filters the problems it is confronting through the history, the beliefs, the language, and the cultural experiences that are seen by its leaders as most likely to engender widespread sympathy and involvement. Understood in this way, framing can be thought of as a device used by the leaders of a movement to provide new understandings of problems that a movement is confronting in a language that is most likely to resonate with particular groups.

The framing of a political movement has several overarching elements, each of which tend to be found in the literature of a movement. In its most basic form, framing first and foremost involves the identification of a problem.[7] Second, it can include what are described by the leaders in a movement as the causes of the problem or problems being confronted by their movement as well as the severity of these problems and the urgent need for dealing with them (Mooney and Hunt 1996, 177–197; Benford 1993a, 201–208).[8] It typically places emphasis on a particular explanation or understanding of a problem, thereby downplaying the significance of other explanations so as to appeal to particular groups.[9] Third, framing can involve extending the organizing of an existing movement or group to encompass problems that are not immediately associated with or connected to the group's work (Snow, Rochford, Worden and Benford 1986, 464–481). Finally, framing is understood here to have a transformative component; it challenges existing understandings or creates new understandings of problems. Not only is a new language for these problems often fashioned in a movement, but a new sense of self-awareness among movement participants (or what is referred to in this book as a collective identity) is also fashioned, creating new attachments among those active in a movement.[10]

The framing in a movement is likened by some scholars to a dramaturgical or a performative process (Benford and Hunt 1992, 36–55; Snow 1979, 23–44). Movements are seen as having a performative or dramaturgical element, insofar as leaders are involved in "scripting," or delineating the identities or roles of participants, antagonists, and bystanders; "staging," or organizing various incidents, protests, and related events to draw attention to their concerns; and "narrating," or providing interpretive accounts of the events of a political movement.[11] It is this kind of dramaturgical perspective that is bringing to the fore the study of political movements' roles in the social construction and contention of political meaning, language, and rhetoric.

The leaders of the SouthWest Organizing Project, the Gulf Coast Tenants Organization, the Oil, Chemical and Atomic Workers Local 4-620, and the Labor/Community Strategy Center have used a language of race, civil rights, and social justice through which to connect hazardous wastes, pollution, ground water contamination, housing conditions, mass transit, local infrastructure, and the related problems of these communities. Through their language of race, rights, and justice, it is postulated that the leaders of these groups have framed their own understanding of the environment, which has more in common with the history of the civil rights, farmworkers, tenants, and other movements than it does with the postwar environmental movement. The redefinition of the environment by civil rights groups and labor unions will be shown in this book to reflect the close relationship between language, culture, and activism in the framing of a political movement.

NOTES

1. "Traditional civil rights organizations," according to Charles Lee of the United Church of Christ Commission for Racial Justice, "have long viewed environmentalism with distrust and hostility" (Lee 1992a, 20).

2. This "optimistic position" that "the growth of environmentalism in rich countries is explained mainly by a post-1968 shift to postmaterialist culture values," concludes Joan Martinez-Alier of the Autonomous University of Barcelona, "only accounts for one variety of environmentalism" (Martinez-Alier 1998, 314–315). Describing what is characterized as an environmentalism of the poor in Latin America and elsewhere, Martinez-Alier concludes that "there are even more reasons to become an environmentalist in poor countries or in poor regions whose environmental space is being used for the benefit of the rich" (Martinez-Alier 1998, 315).

3. Lester Milbrath's pathbreaking study of the postwar environmental movement, for instance, concludes that "many blacks are simply indifferent" to the environment (Milbrath 1984, 16).

4. Most scholarship has underscored the importance of prior organizations as underpinnings of successful movements. What is less appreciated is the extent to which these already established organizations provide important cultural resources as well. As Doug McAdam writes, "what is too often overlooked in structural accounts of [movements] is the extent to which established organizations are themselves embedded in longstanding activist subcultures capable of sustaining the ideational traditions needed to revitalize activism" (McAdam 1994, 43). As McAdam concludes, "these enduring activist subcultures function as repositories of cultural materials into which succeeding generations of activists can dip to fashion ideologically similar but chronologically separate movements" (McAdam 1994, 43).

5. In this sense, writes Doug McAdam, "framing can be thought of as acts of cultural appropriation with movement leaders seeking to tap highly resonant ideational strains in mainstream society (or in a particular target subculture) as a way of galvanizing activism" (McAdam 1994, 37–38).

6. Framing, writes David Snow and Robert Benford, is "an interpretive [scheme] that simplifies and condenses the 'world out there' by selectively punctuating and encoding objects, situations, events, experiences, and sequences of actions within one's present or past environments" (Snow and Benford 1992, 137).

7. At its most basic level, writes Robert D. Benford, framing "involves the diagnosis of some situation or aspect of social life as problematic and in need of change" (Benford 1993a, 1999).

8. As Robert D. Benford writes, there are differences within a movement between simply identifying a problem and convincing persons that it is so severe or urgent that action must be taken. "Just because people agree with a movement's contention that a problem exists," writes Benford, "does not guarantee that they

will drop everything else in their lives and work on alleviating the problem" (Benford 1993a, 201). "Unless the expected undesirable consequences are believed to be immediately forthcoming," concludes Benford, "rationales for postponing action can easily be reconciled" (Benford 1993a, 203).

9. In his work, Robert M. Entman characterizes framing in the following way. "Framing essentially involves selection and salience. To frame is to select some aspects of a perceived reality and make them more salient in a communicating text, in such a way as to promote a particular problem definition, causal interpretation, moral evaluation, and/or treatment recommendation for the item described" (Entman 1993, 52). "Frames, then, define problems, [they] determine what a causal agent is doing with what costs and benefits, usually measured in terms of common cultural values, identify the forces creating the problem, make moral judgments, evaluate causal agents and their effects, and suggest remedies" (Entman 1993, 52). Entman defines salience as "making a piece of information more noticeable, meaningful, or memorable to audiences" (Entman 1993, 53). This is accomplished "by placement or repetition, or by associating them [problems or experiences] with culturally familiar symbols" (Entman 1993, 53). In addition, frames can "direct attention away from other [problems or experiences]" (Entman 1993, 54).

10. People, writes sociologist Pamela E. Oliver, "are changed by social movements. Active participants in collective action are changed by their participation and experiences. The experience of participating often makes them feel more efficacious and ready for more. They often undergo a process of progressive commitment. This process seems to leave them permanently different" (Oliver 1989, 17). As Oliver concludes, "an important part of what happens in a social movement is [that] people change their feelings about themselves" (Oliver 1989, 23).

11. According to Richard Couto in his writing on the civil rights movement, the narrative element in a political movement is especially important (Couto 1993).

2

Race, Ethnicity and the Politics of the Environmental Justice Movement

The environmental justice movement is widely seen as being responsible for the emerging awareness of the affect of environmental problems on the poor, particularly in African-American, Native American, Latino, and Asian Pacific communities. It is argued in this chapter and in the subsequent chapter on the SouthWest Organizing Project that by connecting the environment with more recognizable struggles for civil rights and social justice, the leaders of these groups have increased the involvement of their communities with the environment.

FRAMING RACE AND ETHNICITY IN THE ENVIRONMENTAL JUSTICE MOVEMENT

The environmental justice movement and its activism in African-American, Native American, Asian Pacific, and Latino communities is prompting a resurgence of scholarly interest in the incidence of environmental problems in these communities, largely inspired by the research conducted by the United Church of Christ Commission for Racial Justice in its 1987 study, *Toxic Wastes and Race in the United States: A National Report on the Racial and Socioeconomic Characteristics of Communities with Hazardous Waste Sites.*

The widening awareness in the past decade of environmental problems as they affect the rural and inner-city poor, particularly African-American, Latino, and Asian Pacific communities, is an important objective of the environmental justice movement. Study after study has shown that the poor bear the burdens not only of hazardous wastes but of water and soil contamination, carbon dioxide emissions, asthma, and lead poisoning, all of which put them at even greater risk for health problems (Krieger, Rowley,

Herman, Avery and Phillips 1993, 88–122; Florini, Krumbhaar and Silbergeld 1990). The connection of the environment with impoverishment is evident to many of those Americans who live in the shadows of industry.

Few previous studies of environmental activism in African-American and Latino communities have studied the way in which activist groups are framing a new definition of the environment, in order to incorporate the problems of communities not otherwise involved with the postwar environmental movement. The environmental justice movement is framing problems such as affordable housing and education so as to incorporate them into a new definition of the environment—"where we live, work and play."

This framing of the environment in a language that relates to the circumstances of African-American and Latino communities is an important part of the environmental justice movement, yet it receives relatively little consideration in most scholarly work. With this in mind, this chapter on the Gulf Coast Tenants Organization and the subsequent chapter on the SouthWest Organizing Project argue that their framing of a new definition of the environment is an indispensable part of these groups' activities.

THE GULF COAST TENANTS ORGANIZATION: THE ENVIRONMENT, HISTORY AND HEALTH IN LOUISIANA

The Gulf Coast Tenants Organization was formed in the early 1980s by housing activists in Louisiana, Mississippi, and Alabama. It worked with the residents in public housing projects for nearly a decade before expanding its work to include the environment. Early in its work in these housing projects, the group connected the environmental problems to a history of racial inequalities as well as to the inadequate living conditions in these housing projects.

The historic connection of Louisiana with the environment has been shaped by more than five decades of industrialization, particularly along the banks of the Mississippi River. The chemical industry came to Louisiana in the beginning of the century, as the industry expanded up the Gulf Coast from the oil fields of Texas to the wetlands along the Mississippi River. The chemical industry, lured to this region by an abundance of oil and natural gas as well as the accessibility of transportation on the Mississippi River, expanded into the Gulf Coast of Louisiana through the late 1940s. It sank its roots into the small towns along the river and is to this day one of the largest industries in the state.

The banks of the Mississippi River from Baton Rouge to New Orleans are home to more than one hundred chemical plants. Most of these are large complexes that sprawl across many acres, often hidden in stands of trees or along the edges of large open fields so that only a few smokestacks are visible. In the more densely populated neighborhoods of Baton Rouge, the flares and storage tanks of the plants tower over the roofs of nearby homes.

The sheer magnitude of Louisiana's environmental problems has received national attention in recent years, particularly the impact of industry along the Mississippi River. Louisiana is the highest ranking state in the nation in terms of the amount of hazardous wastes produced per person. The region from Baton Rouge to New Orleans is estimated to generate nearly one eighth of the hazardous wastes in the entire nation (Costner and Thornton 1989). Chemical plants and oil refineries together discharge millions of pounds of wastes into the river each year.

Approximately one and a half million of Louisiana's residents live downstream from this industrialized stretch of the Mississippi River. Contamination of the river is thought by many residents to have a connection with the high incidence of cancer in the region. The hundred and fifty mile stretch of the river is, for this reason, referred to by many local residents as Cancer Alley. While the environment is a problem for all residents of Louisiana, it is especially significant in the towns along the Mississippi River, where environmental problems are compounded by poverty, unemployment, and a lack of basic community services.

FROM HOUSING TO THE ENVIRONMENT FOR THE GULF COAST TENANTS ORGANIZATION

The tenants' movement that emerged from the Gulf Coast region of Louisiana and Mississippi in the early 1980s was enmeshed in the history of political organizing earlier in the century. Tenant organizing became an important part of the civil rights movement in the late 1960s. As the center of some of the most bitter and protracted conflicts of the civil rights movement, the Gulf Coast region of Louisiana and Mississippi does not suffer from lack of a history of political organizing.

A movement of public housing residents emerged in the early 1980s to deal with many of the problems in federally subsidized housing projects. For years, the residents of public housing in the Gulf Coast region had suffered from problems with housing administrators, including overcharging for utilities, evictions, and harassment of residents. In some instances, officials had allegedly searched the apartments and evicted those residents involved in organizing their neighbors.

Beginning in 1982, tenants organizing against these conditions, and those wanting to revitalize tenants' organizations in Louisiana sought the help of organizers with histories of activism. In 1983, working with the Southern Organizing Committee for Economic and Social Justice and the Institute for Southern Studies, these organizers helped form a tenants' organization with offices in New Orleans. As its founders later explained, "a major concern of the tenant leaders was saving their homes from the onslaught of President Reagan's proposal to demolish the worst of public housing and sell the rest to private landlords."[1] Early on, tenant leaders were willing to depict their struggles in a language that had unmistakable connections with the broader struggle for civil rights. In a letter to the Housing Authority of East Baton Rouge in 1986 protesting the inspection of public housing and the alleged evictions of tenant organizers, leaders of the Gulf Coast Tenants Organization explained, "You are trying to get us back on [the] plantation and it won't work. We will not stand for it. We will not stand for people coming into our houses unannounced claiming to be inspecting."

The initial successes of the tenants' movement were the result of a series of protests in Jefferson Parish in 1984. Tenants engaged in an eleven-month campaign to appoint community residents to their local housing board. Following months of rent strikes, demonstrations, and petitions against housing officials, tenants won nearly two million dollars in renovations. Eventually, tenants in the Jefferson Parish Housing Authority also won the appointment of the first tenant in the nation's history to serve as the executive director of a public housing facility. Emboldened by its success, the group organized the residents in federally subsidized projects throughout the Gulf Coast region.

Much of the organization's work was far removed from the environmental problems the group would confront in the subsequent decade. With so many problems—ranging from rent and overcharging utilities to the living conditions in public housing—the work of the tenants' movement was for the most part limited to problems in the housing projects. Although this work was specifically focused on housing, the movement provided experience for organizers who would later focus their work on the environment in these same communities. It was from this success that organizers were also able to widen the issues they confronted in their work, initially focusing on military spending. As tenant leaders later explained, "more than organizing to save their housing, we developed tenant leaders' understanding that to adequately fund their communities' needs the military budget must be deeply cut" (Gulf Coast Tenants Leadership Development Project 1989).[2] In 1984, the nationally known Houses Not Bombs campaign was built from the tenants' groups in the Gulf Coast region (Novotny 1997, 164–166). In its demands for cutbacks in military spending, the Gulf Coast Tenants Organization demonstrated early on that it sought a broader approach to the problems of residents of public housing.

The early involvement of the Gulf Coast Tenants Organization with the environment was a continuation of its work with tenant organizing. Residents in public housing had fought for years against lead poisoning. Some local residents had also been concerned about contaminated drinking water. The group learned from tenants in St. Charles Parish and elsewhere that residents attributed many of their health problems to the nearby chemical plants. For some of the public housing residents who worked with the Gulf Coast Tenants Organization, then, there was an early awareness of the environment.

Many of the residents in public housing in Louisiana also became concerned with the accidents that occurred in the industrial plants along the Mississippi River. In early 1984, an explosion at a chemical plant in Hahnville, Louisiana became a watershed in drawing in the tenants' movement to work on the environment. The explosion led to the evacuation of a public housing project located less than a mile from a plant in Hahnville. Through its newsletters, the Gulf Coast Tenants Organization drew attention to the Hahnville incident and other accidents.

The accomplishments of the tenants' movement in improving the conditions of public housing in places such as Jefferson Parish, coupled with incidents such as the Hahnville plant explosion, led the leaders of the Gulf Coast Tenants Organization to begin to deal with environmental issues. "We taught tenant leaders [the] state and federal environmental laws and regulations, much as we had taught them housing rules and regulations" (Gulf Coast Tenants Leadership Development Project 1989). Tenant leaders learned that local residents had lived with discolored water in their bathtubs and sinks for years. Those who lived near the plants also complained that the airborne pollutants would corrode new wire screens bought for the doors and windows of their homes in a matter of weeks. And for years, the nauseating smells from the plants had been a part of the daily lives of many public housing residents.

The emerging awareness of the environment in the work of the Gulf Coast Tenants Organization occurred at the same time that regional efforts in the early 1980s had begun to draw attention to the connections between racism, poverty, and the environment. In November 1983, tenant leaders attended a gathering of the Urban Environment Conference in New Orleans. The tenants' association also attended the Southeast Environmental Assembly in Atlanta in 1988 (Gulf Coast Tenants Leadership Development Project 1989).[3] Its leaders held a caucus that was addressed by Reverend Benjamin F. Chavis of the United Church of Christ Commission for Racial Justice.

With the leadership of Reverend Benjamin F. Chavis, the United Church of Christ Commission for Racial Justice was crucial to the involvement of the Gulf Coast Tenants Organization with the environment. The publication of the organization's 1987 study, *Toxic Wastes and Race in the United States*, had an impact on the framing of the environment in terms of civil rights and social justice, giving further credibility to what this organization was describing in its own communities as environmental racism. All of these events as well as the publication of the study by the United Church of Christ were significant for the organization's framing of the environment.

THE GULF COAST TENANTS ORGANIZATION AND THE FRAMING OF ENVIRONMENTALISM

The Gulf Coast Tenants Organization emphasizes the history of the civil rights movement in its framing of the environment. This is done by embedding the environment in the familiar experiences of racial discrimination in the Gulf Coast region. Long-standing histories of struggle and of hardship, particularly in the plantations along the Mississippi River, are central to the organization's framing of the environment.

The organization maintains a network of more than fifty tenant associations, which it works with on a regular basis. It also works with any number of neighborhood associations. As in so many other movements in the South, churches also play a vital role in the work of the organization. The leading figures in the tenants' movement regularly speak before church congregations. The organization has also developed a curriculum for what it calls Christian environmental stewardship. In its work with communities along the Mississippi River, the group explains, "Brothers and Sisters, God is on our side. All things can be done if we do His will and trust in Him and each other. God has led our organizations in battle with many of the companies and the politicians they buy" (Gulf Coast Tenants Organization 1993b). "We must preach the gospel of environmental justice," wrote the organization's Pat Bryant in a 1993 pamphlet, "to help religious communities develop the spirituality that has been ours for centuries." The pollution in Louisiana is described by the group as "a national disaster and a spiritual disgrace" (Pat Bryant 1993b). It is a powerful way of framing the environment in communities that are deeply religious and that have turned to their faith as a wellspring for political involvement.

Through its experiences in tenant organizing, the organization has assembled a program of environmental education for some of the poorest residents. The African-American Environmental Education Program is a seven-week course taught by tenant leaders. It includes workshops on environmental health along with readings on the history of the civil rights movement. In the introduction to its Environmental Education Program, executive director Pat Bryant reflects that the staff of the organization "seeks to assist community leaders with what they need and want to know about how to protect the air, land and water where we live, work and play" (Gulf Coast Tenants Organization 1993b). The program is designed to relate to the history of the Gulf Coast region. This is particularly important in the towns along the Mississippi River, where impoverishment makes the conditions for political involvement especially difficult.

THE HISTORY OF LOUISIANA IN THE FRAMING OF ENVIRONMENTALISM

The Gulf Coast Tenants Organization has taken some of the most important experiences of African-Americans in the history of the South and used them as a way of framing the environment. In its work along the Mississippi River, the organization emphasizes that environmental problems are a manifestation of decades of racism. This connects with the historical experiences of many African-Americans, for whom the most evident reminders of this history are the former plantations, whose proximity to the chemical industry is used by the organization to frame the environment.

In terms of African-American history, the Gulf Coast region is among the most important places in the South, with many of the small towns along the Mississippi River dating back to the days of the plantations. With this in mind, the group frames the environment of the region as connected with a history of struggle against racial discrimination. The organization explains that "the goal of the environmental justice movement is to stamp out racism period, racism in housing, the environment, employment, education. These are all tied together. The only way to obtain justice in our environment is to overcome white domination and control over all aspects of our lives" (Gulf Coast Tenants Organization 1993b). Its publications, newsletters, and literature embed the environment in some of the most enduring experiences of African-Americans in the South, particularly life in the former plantations, which remain an evocative metaphor of the history of the region.

The environment is situated by the Gulf Coast Tenants Organization through the experiences of racial discrimination. Describing the chemical companies as "chemical barons" and "environmental carpetbaggers," tenant leaders explain that the residents of public housing have faced even "tougher resistance" against the pollution of their homes and their neighborhoods than against earlier discrimination. As tenant leaders said of a town along the Mississippi River forced to relocate in the face of environmental problems, "the community survived flooding from the Mississippi and it survived Jim Crow, but it could not survive Dow Chemical" (Wright, Bryant and Bullard 1993). The Gulf Coast Tenants Organization was also willing to liken the state of Louisiana and the towns along the Mississippi River to a "Third World" country being exploited by the oil industry.[4]

The environmental problems of the region are likened by the organization to the racism that these communities have had to face in the past. As Janice Dickerson, a former organizer with the organization, explained, "there is no difference from a petrochemical industry located two or three hundred feet from my house and killing me off than there is when the Ku Klux Klan was on the rampage running into black neighborhoods and hanging black people at will" (Di Chiro 1992). The organization taps into this history to frame its definition of the environment to include the inequalities that many of these communities have had to endure, which stretch back, according to tenant leaders, to the era of the plantations.

Because many of the chemical plants are built on the former landholdings of the plantations along the Mississippi River, the connection between the environment and the racial inequalities of the region is common-place in the framing of environmental problems by the Gulf Coast Tenants Organization. The industry is described in the literature of the organization as the "new plantation system" (Wright, Bryant and Bullard 1993). The conditions in the towns surrounding the plants are likened to "the ravages of plantation life" and "the old days of the plantations."[5] As Pat Bryant remarked at the First National People of Color Environmental Leadership Summit in October 1991, "wherever there were communities built by slaves and former slaves, these companies came down and located right beside them. These are all towns that came up from slave communities with petrochemical companies sitting right beside them" (Pat Bryant 1992a). "More than a century old slave communities are being uprooted and a culture that has helped keep African-Americans in this region healthy is threatened" (Gulf Coast Tenants Organization 1992). According to the tenants' group, the chemical industry along the Mississippi River is similar to the plantation system in many ways.

According to its leaders, the Gulf Coast Tenants Organization is able to work in these communities effectively, because it recognizes the frustration felt by many residents over receiving few of the benefits of the chemical industry. Even with the plants and refineries, unemployment in some of the towns along the Mississippi River is more than three times above the average in the state of Louisiana. The Gulf Coast Tenants Organization challenges what it refers to as the "apartheid in employment" in the industry.[6] According to leaders of the tenants' group, it is this perception that the chemical industry has done little to reduce the impoverishment in the region, coupled with what many of these residents see as the adverse health effects stemming from industry, that have led so many residents to become involved with the environment in towns where they would not otherwise be likely to do so.

Much of the focus of the organization has been to bring together these concerns for the environment with struggles to confront impoverishment and a lack of economic development. The industrial property tax exemption in Louisiana, a cornerstone of the industrialization along the Mississippi River that exempts an estimated two billion dollars worth of property annually from state taxes, means that these towns have few tax revenues from the industry and therefore have municipal infrastructures with few public services. "The chemical barons," explains the organization in one of its publications, "have forced these communities to choose between jobs, the companies' sharing the local tax burden, and the communities' health" (Gulf Coast Tenants Leadership Development Project 1989). To this end, the organization has formed a municipal incorporation project to address the lack of infrastructure in the region. The lack of infrastructures in these towns is again likened to "a new form of slavery."

THE CIVIL RIGHTS MOVEMENT AND THE FRAMING OF ENVIRONMENTALISM BY THE GULF COAST TENANTS ORGANIZATION

The Gulf Coast Tenants Organization's framing of the environment is dominated by the history of the civil rights movement. The group reminds its members of the continuity of its struggle against environmental problems today with the struggles of the civil rights movement. In his address to the First National People of Color Environmental Leadership Summit in October 1991, Pat Bryant reflected at length on this. "Our struggles did not just begin. The environmental struggles did not just begin six years ago when the Gulf Coast Tenants Organization began to look at the poisoning of people in Cancer Alley. The environmental struggle did not just begin ten years ago when the Southern Organizing Committee for Economic and Social Justice helped us to begin the Gulf Coast Tenants Organization. It did not just begin in the freedom movement of the 1960s nor with the Student Nonviolent Coordinating Committee, the Southern Christian Leadership Conference, and the Southern Negro Youth Conference" (Pat Bryant 1992a). The civil rights, labor, and tenants' movements have all been influential in shaping the work of the tenants' organization. Even in its flyers and leaflets, the organization's leaders invoke the history of the civil rights movement.

While the framing of these environmental problems as environmental racism has been important for the organization, its march along the Mississippi River in November 1988 drew national attention to the group and reinforced the connections between the civil rights movement and the environmental justice movement. To a large extent, it was this eleven-day march from New Orleans to Baton Rouge that did more than anything up to that point to connect the organizing of African-Americans against environmental problems with the history of the civil rights movement.

As the 1988 march of the Gulf Coast Tenants Organization made its way through the towns along the Mississippi River that are home to the chemical plants, it challenged the industry by invoking the struggles of the civil rights movement.[7] The march was also a watershed in that it coincided with the publication by the federal government of the first comprehensive documentation of hazardous emissions in Louisiana and throughout the nation. The U.S. Environmental Protection Agency's Toxic Release Inventory had been published just prior to the march under the provisions of the 1986 Superfund Reauthorization Act. Marchers took advantage of this as they walked through the towns and held press conferences and prayer vigils at each plant to publicize the newly released industrial emissions figures.

The Gulf Coast Tenants Organization has initiated several projects to challenge the postwar environmental movement by connecting its own struggles against the chemical industry with the history of the civil rights movement. Among the most important of these has been a campaign to change Earth Day from April to March, to commemorate the 1965 Bloody Sunday attack on civil rights marchers in Selma, Alabama. The tenants' group launched its Change the Earth Day campaign "in order to make clear the roots from which we come and to build on the strength that this gives us" (Gulf Coast Tenants Organization 1993a). It involves changing the celebration of Earth Day from April 22 to March 7, the day in 1965 when six hundred civil rights marchers were attacked by Alabama state police on the Edmund Pettus Bridge in Selma. The enduring significance of the March 1965 events for the history of the civil rights movement has been used by the Gulf Coast Tenants Organization as a means of framing their challenge to the postwar environmental movement, reminding all those involved that the civil rights movement was still waging some of its fiercest struggles when the environmental movement was exploding onto the scene in the late 1960s.

The Change the Earth Day campaign is the most recent effort in the work of the Gulf Coast Tenants Organization to frame the environment in terms of the history of struggles for racial justice in the South.[8] The Change the Earth Day campaign is one example of the far-reaching changes in the understanding of the environment that with which the organization has been involved, as activists have contested the environmental movement's most important annual event and thereby linked the environment with one of the most historic moments of the civil rights movement.

The history of the civil rights movement has an enduring significance that has shaped the way members of the Gulf Coast Tenants Organization understand the environment. The group is rallying communities to create what it calls "Selmas of the 1990s." "When African-Americans decided that racial segregation would be ended, even if lives had to be lost in the struggle, Jim Crow was killed. Now leaders of people of color in the South, joined by a growing number of working-class whites and other fair-minded and concerned European Americans, are uniting to stop the poisoning of our workplaces and communities" (Gulf Coast Tenants Organization 1993a). This history of the civil rights movement and its legacy of struggle is invoked by the Gulf Coast Tenants Organization, as it is by many other groups in the environmental justice movement.

The Southern Community/Labor Conference for Environmental Justice was sponsored by the organization in December 1992. At this conference, the group circulated the *Southern Action Manifesto for Environmental and Economic Justice* (1993a), which is widely recognized as one of the most important statements to date by the environmental justice movement. In this manifesto, the organization placed its work with the environment in a much wider historic context. "Beyond our fight to destroy environmental racism and injustice, we are uniting for a future that provides adequate housing for everyone at affordable prices, clean air, land and water for all, education without regard to color and class status, safe jobs and decent wages, restoration of our health, and monetary compensation to those who have been poisoned by industry and the government" (Gulf Coast Tenants Organization 1993a). In the view of many of the leaders of the organization, their communities have not had the luxury of organizing around the environment as a single-issue concern.

This manifesto is one of the environmental justice movement's clearest attempts to connect the environment with the history of the civil rights movement. Much of its emphasis is on defining the environment in terms of the communities' understanding of it rather than allowing "the 'environment' to be defined narrowly so that it becomes the arena of a few single-issue, white, male-dominated national organizations and a tool for abuses by profiteering business organizations" (Gulf Coast Tenants Organization 1993a). Such criticisms of the postwar environmental movement provide a distinctive role for African-Americans and other persons of color, whose experiences, the organization argues, make them capable of forming a wider understanding of the environment.

Toward the end of the manifesto, the organization went even further in relating the struggles against pollution in Louisiana and elsewhere in the South to the unfinished struggles of the civil rights movement. Connecting with an earlier history of civil rights struggles against literacy tests and poll taxes, the organization's statement urged efforts to register more than six million new voters in the South. "We are alarmed by the fact that since the passage of the Voting Rights Act and the dramatic increase in African-American and Latino elected officials in the South, there has also been a dramatic increase in the poisoning of our communities. We are determined to campaign for and elect officials who will be accountable to us and will fight for environmental justice."[9] The statement concluded with a call for nonviolent civil disobedience to draw attention to the environment in the South, in much the same way that the March 1965 events in Selma awakened the nation to the civil rights movement, a point reiterated by the Gulf Coast Tenants Organization as part of its Change the Earth Day campaign.

THE FRAMING OF THE POSTWAR ENVIRONMENTAL MOVEMENT BY THE GULF COAST TENANTS ORGANIZATION

The Gulf Coast Tenants Organization remains unyielding in its criticism of the nation's largest environmental organizations. The environmental movement is viewed as remaining "conspicuously silent" regarding the history of racial inequalities in the South. It is depicted as a movement bound up in a history of "racial inequalities that make it very difficult to build a movement with the power to change the conditions of our poisoning" (Wright, Bryant and Bullard 1993). The leaders of the Gulf Coast Tenants Organization have framed the lack of racial and cultural diversity among the environmental movement's leading organizations as the "Achilles heel" of the movement.

In January 1990, the Gulf Coast Tenants Organization circulated its much-publicized letter to the nation's largest environmental organizations. It was sent to the executive directors of the National Wildlife Federation, the Sierra Club, the National Audubon Society, the Environmental Defense Fund, the Friends of the Earth, the Izaak Walton League, the National Parks and Conservation Association, and the Natural Resources Defense Fund. The letter challenged what the tenants' group referred to as the "institutional racism" of these organizations. It also brought attention to the work of the Gulf Coast Tenants Organization, particularly when it was featured by *The New York Times* several weeks before the twentieth-anniversary events of Earth Day. The letter also coincided with an April 1990 tenants' march along the Mississippi River.

In what is some of its sharpest criticism of the postwar environmental movement, the Gulf Coast Tenants Organization has argued that the nation's environmental organizations are in many instances less diverse in the hiring, employment, and promotion of people of color than the industries that are seen as responsible for the pollution in Louisiana. As an influential figure in the organization has written, "when they roll in to defend the world against poison mongers, minority communities can hardly miss the fact that environmental groups have less inclusive affirmative action policies than the toxic-producing companies themselves" (Pat Bryant 1989). On the occasion of its April 1990 march along the Mississippi River, the group explained that "the nation's large, national environmental organizations are *part of the problem*. They have been silent on the poisoning of Third World people. And their hiring records are *worse than some of the plants that poison us*. They must change now by hiring leaders from the communities where people of color are fighting poisons. If they don't, they must be stopped [emphasis added]" (Gulf Coast Tenants Organization 1990b). In this forcefully worded statement, the Gulf Coast Tenants Organization reflects the frustration that many in the environmental justice movement feel toward the nation's largest environmental organizations. Most of these organizations continue to be regarded as indifferent to local residents, even though groups such as Greenpeace and the former National Toxics Campaign have worked closely with the Gulf Coast Tenants Organization on the environment along the Mississippi River.

What is widely characterized (in the literature and publications of the group) as a narrow understanding of the environment has done much to limit the involvement of poor and predominantly African-American communities with the postwar environmental movement, according to the leaders of the Gulf Coast Tenants Organization. Most of the literature of the group emphasizes that the nation's largest environmental groups have largely ignored these communities, even though elsewhere in its literature the group acknowledges that the Sierra Club has been involved with the environmental problems along the Mississippi River. Although environmental organizations have largely been unable to organize in impoverished communities, groups such as the Gulf Coast Tenants Organization, which have worked for years on tenants' problems and other community issues, *are* able to organize with the environment in these communities. This success is due, to a large extent, to their framing of the environment as having a connection with their previous organizing and with the many different concerns of local residents.

REDEFINING THE ENVIRONMENT BY THE GULF COAST TENANTS ORGANIZATION

The Gulf Coast Tenants Organization has brought together different issues in its work with the environment. The group has widened its definition of the environment to include the problems of poverty, racism, and a history of political disenfranchisement. Its framing of the environment has also included problems ranging from homelessness and police brutality to unemployment and education. The organization brings a history of organizing for civil rights to its work with the environment that few other groups could tap into in the towns and public housing projects along the Mississippi River.

The widely known understanding of the environment as "where we live, work and play" is emphasized by the Gulf Coast Tenants Organization as a way of subsuming many of the problems previously dealt with by the group. Environmental issues are framed as problems of "racial and economic justice." In this way, its new definition of the environment is understood as a part of "all the other factors that affect where we live and work such as housing, health care and economic development" (Gulf Coast Tenants Organization 1993b). Its new understanding of the environment as "where we live, work and play" subsumes nearly all of the problems the organization has previously worked on, so that there is less discordance between its organizing in public housing projects and its work with the environment.

In its literature, the organization emphasizes its "new definition" of the environment. What this means is that the group now has an understanding of the environment that can include nearly any problem that the group has worked with in the past. "We [have] had to learn and to teach a new discipline, environmental protection, and weave it into the real world where our leaders live, work and play" (Gulf Coast Tenants Organization 1992). Indeed, the Principles of Unity statement from its November 1988 march along the Mississippi River was titled "Redefinition of the Environment," reflecting the framing of the environment and its importance in the work of the organization.

The environment, affordable housing, and unemployment are seen as connected with one another. In a statement released with the Southern Organizing Committee for Economic and Social Justice, the organization explained that "the environmental justice movement in the South has developed from the grassroots in communities across the region and inherently incorporates all the other life-and-death issues its activists confront: joblessness, abusive police practices, and the lack of health care, decent housing, and equitable education. Thus, it bases itself on the new definition of the 'environment' which includes all the life conditions where we live, work and play" (Gulf Coast Tenants Organization and Southern Organizing Committee for Economic and Social Justice 1993). The group is sensitive to the impact that the language used to describe the environment can have in terms of how it is understood by their communities.

The organization has extended its work much farther than tenants' and housing problems. Many of the residents in these communities had lived with these problems for years, according to leaders of the tenants' group. However, the group's new definition of the environment as "where we live, work and play" is creating an opportunity to involve those who might not otherwise understand the hazards they face in their communities or regard the pollutants discharged into the Mississippi River as environmental problems.

The leaders of the organization acknowledge that many other problems take precedence over the environment in the public housing projects, where the group began organizing more than a decade ago. These housing projects—in New Orleans, Baton Rouge, and elsewhere—still have to deal with problems of substandard living conditions along with crime, drug abuse, school suspensions, unemployment, and a litany of other concerns. With the evictions, poverty, and the many other problems in these housing projects, "the chemical dangers pale in contrast to the rest" (Gulf Coast Tenants Organization 1993d). Instead of letting this limit its work with the environment, however, the group continues to connect the environment with these many other problems. "We include this long-range focus [on the environment] with the more narrowly defined day-to-day targets of adequate maintenance, tenant participation in management decisions, and eviction relief," explains the group (Gulf Coast Tenants Leadership Development Project 1989). Its struggles against rent and utilities overcharging continue to be interwoven into the larger struggle for social and environmental justice. Far from presuming the environment to be something that is "irrelevant" to their communities, the organization has recast its work with the environment to incorporate these seemingly disparate problems.

The organization has been effective in its work with the communities along the Mississippi River because of to its involvement with other concerns as well as its appropriation of powerfully resonant cultural symbols for its framing of the environment, in particular, symbols of racism, social justice and civil rights as well as a history of the plantations and even the religious faith that is a cornerstone for so many African-American communities. The group has built a reputation that allows it to work with the environment in communities that would otherwise not be likely to be involved with environmental issues.

From its involvement with the environmental problems of the public housing projects and the towns along the Mississippi River, the organization has emerged as a leading organization in the environmental justice movement. With its history in the civil rights and tenants' movements, the organization is uniquely positioned to form a new kind of environmentalism that has more in common with these histories. From its framing of the chemical plants as the "new plantations" to its drawing from religious imagery in depicting the struggles of community residents, the organization grounds its work in a history of African-American struggles against racial discrimination which, in turn, is leading to a unique kind of environmentalism in communities which had little prior connection to the environmental movement.

NOTES

1. As Pat Bryant wrote in 1982, "the long range objective of the [Reagan] administration is to sell public housing to private real estate developers (who charge exorbitant rent and usually won't rent at all to large, especially black, families), or demolish it to make room for commercial investments" (Bryant 1982, 29). Organizers with the Gulf Coast Tenants Organization were able to organize local tenant leaders with charges in their literature such as, "Reagan's plan [is] to replace public housing with subsidies to private landlords."

2. Explained Pat Bryant in 1982, "many tenants are becoming convinced that public housing will survive only if the nation's priorities are changed, from guns to housing to nutrition, health care and education" (Bryant 1982, 31).

3. Among other groups attending the 1988 Atlanta conference was the SouthWest Organizing Project from Albuquerque.

4. Wrote Darryl Malek-Wiley, an organizer of the 1988 march along the Mississippi River, "enormous economic benefits have accrued to the oil industry while the people of Louisiana have been and are now forced to pay a terrible price in damage to the environment and the quality of life. The oil industry and its willing accomplices in state and local government have treated Louisiana like a Third World country by extracting the state's mineral wealth and funneling most of the profits to out-of-state corporations and banks" (Malek-Wiley 1988, 2).

5. As one participant at the Southern Community/Labor Conference for Environmental Justice in December 1992 explained, "an old plantation house that is now the company headquarters, once devoted to sugar, now is devoted to the petrochemical industry. The people affected most are not [those in] the big house but the people in the surrounding slave and sharecropper quarters." Another conference participant explained, "it has a nasty and weird continuity, plantation big houses, gorgeous but symbols of oppression and misery, the very sites of the

manor houses are now the locations of the offices of these chemical and fertilizer companies spewing out a different kind of oppression and poison" (Gulf Coast Tenants Organization 1993b, 157).

6. In *Louisiana Great Toxics March Principles of Unity*, the tenants' group reflected "the petrochemical plants should stop their longstanding policy of hiring workers from outside the local community that has to bear the social and environmental costs of the petrochemical plants being located there. Instead, these plants should initiate a hiring program of affirmative action, to bring more local workers, particularly ethnic workers, into the plants at all levels, including management positions, not only the low-paying jobs" (Gulf Coast Tenants Organization 1990a).

7. Writing in the days prior to the 1988 march along the Mississippi River, Darryl Malek-Wiley stated that "environmental issues will be linked [by the march] to the erosion of the area's cultural heritage and family life, the destruction of historical buildings, and the alarming health consequences facing the region's workers and residents" (Malek-Wiley 1988, 1–2).

8. Pat Bryant of the Gulf Coast Tenants Organization has explained "we should have the audacity to rename and reclaim something that has been taken from us. Dennis Hayes and all his colleagues, in good faith, took on and built Earth Day. Let's take it back." Pat Bryant, Gulf Coast Tenants Organization, in United Church of Christ Commission 1992, 132. Bryant's remarks were made at the First National People of Color Environmental Leadership Summit in October 1991. Dennis Hayes was one of the leading organizers of the Earth Day activities in April 1990.

9. As the Gulf Coast Tenants Organization explained in *A Short History of the Gulf Coast Tenants Organization* (1992), "state and federal agencies responsible for protecting the environment look the other way as the poisoning continues. Some of our politicians are the best that money can buy. That is to say, petrochemical and chemical waste industry PAC money has been the big factor in Louisiana elections for some time. And it is becoming a more important factor in Mississippi and Alabama."

3

The SouthWest Organizing Project

The SouthWest Organizing Project is a leading group in the environmental justice movement. It has been effective in its work in Albuquerque and throughout New Mexico because of its long involvement with other concerns in these communities as well as its use of cultural and historical narratives for its framing of the environment. In a way similar to the Coast Tenants Organization and its framing of the environment, the SouthWest Organizing Project makes a connection between the colonization in the eighteenth-century and the environment today in New Mexico and throughout the Southwest.

ENVIRONMENT, LAND AND POLLUTION IN THE SOUTHWEST

Any understanding of the environment in the Southwest is rooted in the centuries of economic development in the region. In the nineteenth century, the environment was transformed by the expansion of cattle ranchers and homesteaders into what is today New Mexico and Arizona. This was facilitated by the railroads at the end of the nineteenth-century. Commercial logging in the mountains also led to deforestation in some parts of the region.

The commercial expansion in New Mexico and throughout the Southwest, as well as the expansion of railroads, led to overgrazing, erosion of the soil, and the destruction of farmland that had sustained communities for centuries (Forrest 1989). In the late 1880s, the largest ranches continued to expand their landholdings, displacing communities that had farmed the land in this part of the country for centuries.

Although there is a long history of environmental degradation in the Southwest that dates back hundreds of years, many of the problems being faced in the region today are a result of more recent changes, during the postwar period. The environment of the Southwest, writes Mike Davis, has been transformed by "militarism, urbanization, an interstate highway system, mass tourism, and the extractive industries' boom-and-bust cycles" (Davis 1993, 58). The economic growth in the Southwest in the past fifty years has left a legacy of environmental despoliation that has changed the lives of communities throughout the region. Because of the damage caused by industrialization and real estate development, local residents have likened the Southwest to a national dumping ground.

A history of colonization in the Southwest shapes the framing of the environment by the SouthWest Organizing Project in much the same way that a history of racial inequality shapes the framing of the environment by the Gulf Coast Tenants Organization. The environment, for the SouthWest Organizing Project, is tied to the encroachment of Anglo culture in the region. Its literature is replete with references to what it calls a history of colonialism in the Southwest.

Environmental problems are numerous in New Mexico, whose arid climate is especially vulnerable to surface and groundwater contamination. Large portions of New Mexico's surface water have been contaminated by mining and agriculture in the state (Hall and Kerr 1991). New Mexico also has the largest number of military-related waste sites in the nation. These have had an even greater impact on the environment than have many of the state's industries.

FROM STUDENT MOVEMENT TO COMMUNITY ORGANIZING: A HISTORY OF THE SOUTHWEST ORGANIZING PROJECT

The SouthWest Organizing Project has a history of organizing on a wide array of issues. In the early 1980s, welfare and tenant activists in Albuquerque sought to challenge the Reagan administration, particularly its cutbacks in housing and welfare programs. Much like the Gulf Coast Tenants Organization in Louisiana, its organizers were determined to form a group that would build coalitions around what it saw as some of the most intractable social problems of the day.

"To empower the disenfranchised in the Southwest to realize racial and gender equality, and social and economic justice," reads the organization's mission statement, reflecting the many different issues addressed by the SouthWest Organizing Project. "At the SouthWest Organizing Project," according to the masthead from the group's publication *Unidas Voces*, "we are working for the self-determination of all peoples." The group was founded in 1981 by activists who envisioned an organization that would deal with many different issues. It has worked with a host of issues including affordable housing, neighborhood economic development, and voter registration, signing up almost twenty thousand voters in Albuquerque by 1984.

When it was first established, the SouthWest Organizing Project was closely identified with the university protests of the early 1970s. This period had witnessed a flowering of activism, much of it centered in the neighborhoods of Albuquerque's South Valley. The murals, theater, poetry, and writing that flourished in the early 1970s were crucial to the project.

Even though environmental problems were far removed from the work of the group in its early years, the environment was nothing new to its founders. In the early 1970s, the issue of the environment percolated through the student movement, thanks in part to a renewed interest in the history of the land in New Mexico and elsewhere in the Southwest. The Brown Berets and United Farm Workers had a nascent environmental consciousness, which the founders of the SouthWest Organizing Project drew upon when they later turned their attention to the environment.

Local residents in Albuquerque's South Valley began to work with the environment in the 1970s. They expressed concern about a municipal sewerage treatment plant in their community. These local residents, who included some of the founders of the SouthWest Organizing Project, became concerned enough to organize a protest in conjunction with Earth Day in April 1970. The Chicano Earth Day drew more than three hundred participants.[1] Marchers, including students from the University of New Mexico, held a protest outside the treatment plant.[2] From this, organizers learned first-hand of the possibilities for organizing around the issue of the environment.

Following local protests against the sewerage treatment plant in the South Valley, the organizers who would found the SouthWest Organizing Project turned to a nearby particle board factory. Local residents became involved with the issue of the factory's accountability to the community as well as its health threats. The project's initial door-to-door canvassing made organizers attentive to the ways in which poor and working-class communities understand the environment.

Early in its work in 1981, the SouthWest Organizing Project learned that for years, families in the South Valley had complained of smoke from the particle board factory. Residents had also complained of groundwater contamination. Many of them, however, were unlikely to think of these issues in terms of the environment. What organizers with the project learned was that the community uses language in its own way to express its concern with the environment. Louis Head and Michael Guerrero later reflected that "residents [of the South Valley] were not very responsive to questions concerning the 'environment' or 'environmental issues.' However, when asked about 'community' issues of concern, people would rank groundwater and air contamination high on the list, often in terms of the failure of state and local governments to do anything about such problems" (Head and Guerrero 1991, 7). Local residents felt that their problems were a reflection of the indifference of local officials toward the residents of the South Valley.[3] In short, their concern with the environment expressed itself in terms such as health, government responsiveness, or community services rather than the environment.

The SouthWest Organizing Project's work in the South Valley led it to look at the environment throughout New Mexico. From its initial canvassing, it found that local residents expressed more interest in the environment when organizers drew connections with problems of unemployment, affordable housing, or local schools. "Based on concerns expressed in surveys and house meetings and on what we saw as national trends, we came to recognize environmental degradation as a social and racial justice issue. A new term, environmental racism, became a part of our language" (SouthWest Organizing Project 1991, 7). The persistence of poverty in the Southwest, coupled with its arid climate, made the environment an especially pressing issue. The SouthWest Organizing Project soon developed a new understanding of the environment as a social and racial justice issue.

ORGANIZING AROUND THE ENVIRONMENT IN THE SOUTHWEST ORGANIZING PROJECT

Early in its work, the SouthWest Organizing Project framed poverty and the environment as interlocking issues, inseparable from housing, workplace, or community problems. Environmental organizations such as the Sierra Club were criticized for what the Albuquerque group characterized as a narrow understanding of the environment, one seen as indifferent to the often difficult realities of the underprivileged in New Mexico.

The constituency of the SouthWest Organizing Project is largely made up of working-class Latinos, many of whom had little involvement with the postwar environmental movement. For this reason, the leaders of the group began to frame the environment to fit with their previous work. "The potential exists for a new understanding to emerge between the [mainstream] environmental movement and organizations working for social and racial justice. The process will require better defining how issues of [the] environment fit within a social, racial and economic justice framework" (Head and Guerrero 1991). The group sees the environment as a unifying issue.

In SouthWest Organizing Project publications, environmental problems and the economic inequality of these communities are all but indistinguishable. "Environmental racism is very real and environmental issues cannot be separated from the broader context of social, racial, political and economic injustice" (Head and Guerrero 1991, 8). Environmental problems "are issues of social and racial justice. Actions that lead to toxic contamination of communities of color are acts of institutional racism if not systematic racial violence" (Moore and Head 1994). "The systematic exposure of people of color to hazardous wastes is one of the most insidious forms of racism today" (Head and Guerrero 1991, 4). The underlying assumption of such actions "is that people of color are expendable" (Head and Guerrero 1991, 4). This understanding of the environment as "systematic racial violence" is central to the group's framing of the environment.

Few aspects of the work of the SouthWest Organizing Project are as well established as its framing of the environment as "issues of social and racial justice." Environmental racism, according to the Albuquerque group, reflects a history of discrimination. This incidence of hazards is seen as caused by the "racism inherent in land-use decisions." "The term gentrification," explains the group in its pamphlet *Development, Environment and Race in New Mexico* (1994), "could have been invented in New Mexico." The SouthWest Organizing Project draws from this history as a way of framing the environment. One aspect of discrimination with which the group is concerned is what it sees as the cultural ramifications of environmental problems. The land and the water of New Mexico are described as "extremely precious cultural resource[s]" and "basic component[s] of cultural survival for people in New Mexico."[4]

The colonial heritage of the Southwest as well as its history of agricultural development are held responsible for many of the region's environmental problems. The history of the land in the Southwest and the modern development of the region are so inextricably bound for the SouthWest Organizing Project as to be practically indistinguishable in the way they frame the environment. As Louis Head and Michael Guerrero write, "two threads of history explain the disproportionate impact of toxics on communities of color. One is the long history of oppression of people of color in the United States, from indentured servitude, slavery and colonization to newer forms of institutionalized racism" (Head and Guerrero 1991, 4–5). In addition, people of color "have historically been drawn into the worst jobs in agribusiness and heavy industry [and] they often end up living in areas of heavy pollution" (Martinez 1991b, 3). The group ties economic development in the state to "environmental degradation and economic injustice," which affects people of color most intensely.

The Albuquerque group seeks what it calls the "broadest context" for its work with the environment. The group is reluctant to turn its undivided attention to protecting the environment without attention to the struggle for places where people can live, study, and work without fear of health or environmental threats. As Richard Moore and Louis Head write, "we are not an environmental organization but rather address toxic contamination as part of a broad set of racial and social justice concerns. We do not single out the environment" (Moore and Head 1994, 191).

The SouthWest Organizing Project "does not consider itself an 'environmental' organization but rather a community-based organization which addresses toxics issues as part of a broader agenda of action to realize social, racial, and economic justice. We do not single out the environment as having a special place above all other issues, rather we recognize that issues of toxic contamination fit within an agenda which can and, in our practical day-to-day work does, include employment, education, housing, health care, and other issues of social, racial and economic justice" (SouthWest Organizing Project 1990b). In this way, the project maintains its identity as a community-based, multi-issue group despite the growing importance of the environment to its work.

The group is also making its outreach to churches a priority. With members of church parishes in Albuquerque, the project surveyed neighborhoods to determine what environmental problems were important to local residents. The group has worked with the *El Buen Samaritano* United Methodist Church to survey Albuquerque's Sawmill, Old Town, and Wells Park neighborhoods.

ENVIRONMENTALISM AND ITS HISTORY IN THE SOUTHWEST

The SouthWest Organizing Project has had to go to some lengths to frame its work as distinct from that of the nation's environmental movement. Its literature tends to be critical of the postwar environmental movement, whose history in the Southwest is interwoven into a historic narrative of the encroachment of Anglo culture on traditional culture in the Southwest.

The postwar environmental movement has a history of what the SouthWest Organizing Project sees as indifference to the concerns of their communities. Although federal land reclamation projects were started early in the century to deal with the erosion caused by the cattle and timber industry in New Mexico, these range management and soil conservation programs also had the effect of displacing some local communities. As a result, activists working with the environment in the Southwest today remain mistrustful of the U.S. Forest Service and other environmental agencies.[5] In some instances, environmental groups and even federal agencies such as the U.S. Environmental Protection Agency are also framed by the project as having been "detrimental" to the interests of the poor and people of color in the Southwest.

On the basis of its work with the environment in the South Valley and elsewhere, the SouthWest Organizing Project has described the nation's largest environmental organizations as the "new colonizers." Environmental groups are for the most part seen as having had little regard in the past for local communities as well as having been indifferent to the immediate issues of survival for poor and working-class Latinos in the Southwest.[6]

The project is unyielding in its criticism of the nation's environmental movement. The group became nationally known when its organizers sent a letter to the nation's largest environmental organizations in March 1990 signed by more than one hundred religious and civil rights leaders. The letter represented a watershed for the SouthWest Organizing Project and marked its emergence as a leading group in the environmental justice movement. It contended, "for centuries, people of color in our region have been subjected to racist and genocidal practices including the theft of lands and water, the murder of innocent people, and the degradation of our environment" (SouthWest Organizing Project 1990a). The letter also accused the nation's environmental groups of having "histories of racist and exclusionary practices" (SouthWest Organizing Project 1990a). "Racism is a root cause of your inaction [in] addressing [the] environmental problems in our communities" (SouthWest Organizing Project 1990a, 3). Coinciding with the Earth Day events in April 1990, the nation's largest environmental organizations were sent scrambling by the challenge of a modestly sized group located in a quiet residential neighborhood of Albuquerque.

The framing of the work of the postwar environmental movement as part of the history of colonization in the region is an important part of the SouthWest Organizing Project's work with the environment. Environmental organizations such as the Sierra Club and the Nature Conservancy are framed as part of a history of commercial expansion and the destruction of the environment in the region. They are also depicted by the SouthWest Organizing Project as linked with major corporations.[7] In likening these organizations to the "new colonizers" and challenging their "institutional racism," the SouthWest Organizing Project has framed its work as distinct from the work of these organizations.

HISTORIES OF RESISTANCE AND THE ENVIRONMENT IN THE SOUTHWEST

The Southwest has a long history of movements for social change that extends back for centuries (Hall 1989). The SouthWest Organizing Project draws from a history of these movements, particularly the farmworkers and student movements of the 1960s and early 1970s. The struggle with the environment is seen as yet another phase in the history of these struggles against Anglo encroachment in the Southwest. In this sense, the environment is bound to the histories of local struggles for justice.

A thread of cultural history has wound its way through nearly every campaign of the SouthWest Organizing Project. This is understandable in the historical context of the Southwest, whose history has been one of the encroachment of lands that have in some instances been handed down in families for generations (Acuna 1972). The resistance against the seizure of lands, the "cutting down of the wires" of the fences that enclosed the large ranches and landholdings, as some members of the group put it, is crucial to what the group calls its "histories of resistance."[8] This history, in turn, is crucial to the group's framing of the environment.

Much like the work of the Gulf Coast Tenants Organization with the towns along the Mississippi River, the SouthWest Organizing Project connects its work with what it sees as a history of colonization of the region. "In the past, our lands, the lands of indigenous people, were stolen from us, and we were poisoned with smallpox infested blankets, our animal relatives slaughtered, and our waters contaminated. This poisoning of communities of color has continued unabated, varying only in intensity and scale" (SouthWest Organizing Project 1993a). Today, the SouthWest Organizing Project criticizes what it calls the "rural land grab" by real estate developers. Displacement from the land, whether by real estate development or reclamation projects, is tantamount to what it calls "cultural genocide," according to the group (SouthWest Organizing Project 1994). The significance of this history of the land is crucial to the group.

The project and its work has drawn from the history of impoverishment and even exploitation in the region. In some of its literature, the group has gone so far as to describe poor and working-class Latinos in the Southwest as the true endangered species of the region. For the past several years, the organizing of the SouthWest Organizing Project against the semiconductor industry has been a particularly important way in which the group has taken a stand against the impact of commercial development on the environment in New Mexico.

THE SEMICONDUCTOR INDUSTRY AND CORPORATE ACCOUNTABILITY FOR THE SOUTHWEST ORGANIZING PROJECT

In the past decade, New Mexico has emerged as a manufacturing center for the semiconductor industry. The largest firm to expand its facilities is Intel Corporation, which has constructed a large assembly facility at Rio Rancho along the Rio Grande River. The SouthWest Organizing Project has been working on a campaign to respond to Intel since April 1993, when the corporation announced a one-billion-dollar expansion of its facilities in Albuquerque. Initially, the project focused on the subsidies granted to the plant by the state of New Mexico. The group has also leveled a charge that the corporation moved its assembly plant from a proposed site in California to avoid that state's environmental and labor regulations.

The SouthWest Organizing Project is leading the challenge against the granting of subsidies to the semiconductor industry. In particular, it alleges the loss of more than one hundred million dollars in revenues for the state of New Mexico due to tax abatements and state-assisted financing for the Intel plant. For all of New Mexico's encouragement of the semiconductor industry, the SouthWest Organizing Project insists that the employment benefits of the infrastructure financing by the state of New Mexico are negligible. In its 1995 study, *Intel Inside New Mexico: A Case Study of Environmental and Economic Injustice*, the group argued that the majority of jobs in the industry are not particularly skilled and pay less than the national average for manufacturing employment. This does not include the effects of the industry on the health of its workers, who often have to deal with hazardous materials.

From the sparsely decorated rooms of its office in Albuquerque, the SouthWest Organizing Project has challenged some of the largest corporations in the semiconductor industry and drawn attention to what it depicts as the health and environmental dangers of the industry.[9] The workforce in the plants is seen as bearing the brunt of health and environmental problems. The Albuquerque group contends that some employees have had to work in poorly ventilated spaces. The expansion of the industry, according to group leaders, will also require millions of gallons of water a day. As a result, many residents have become concerned about contamination of the Rio Grande River. For its part, the project has been attempting to obtain off-site monitoring of air and groundwater emissions in the vicinity of the Rio Rancho facilities.

As part of a history of colonization in the Southwest, the lack of economic development in many of these communities is framed by the project as leaving them vulnerable to environmental problems. New Mexico is depicted in the literature of the group as having lowered its standards of environmental protection to attract industry from outside of the state. Local communities, the project adds, have also been pitted against one another in offering investment subsidies and property tax abatements to industry, further reducing the fiscal resources for already cash-strapped communities.

Even in its work with the semiconductor industry, the language of colonization threads its way through the publications of the SouthWest Organizing Project.[10] Much like the work of the Gulf Coast Tenants Organization and its framing of the environment in the towns along the Mississippi River, the SouthWest Organizing Project seeks to make a connection between the colonization of the eighteenth-century and the environment today in New Mexico, even when taking on the latest high technology firms in the state.

FRAMING THE ENVIRONMENT BY THE SOUTHWEST ORGANIZING PROJECT

The SouthWest Organizing Project, like many groups in the environmental justice movement, is involved in framing the environment as "where we live, work and play." Every movement must persuade its participants and potential participants that one particular problem is more serious than others. The SouthWest Organizing Project has done this by framing the environment as environmental racism, which incorporates many of the problems dealt with by the group. Its definition of the environment as "where we live, work and play" frames it as an issue that brings together many of the other issues in these communities.

The work of the project with the environment is part of a much larger effort to address issues of impoverishment and cultural disempowerment. The environment, according to the literature of the group, is a unifying issue through which to address "the disempowerment that people have in their lives and the work of people to reclaim their communities" (South-

West Organizing Project 1991). As part of this, the group is working with activists in Albuquerque on affordable housing and economic development there, while encouraging local investment in environmentally oriented manufacturing.

The project is connecting the environment with the many more immediately recognizable problems faced by lower income persons in the region. Much of this is done through its new definition of the environment, a definition that is found in the literature, publications, and newsletters of the group. Through its definition of the environment as "where we live, work and play," write Louis Head and Michael Guerrero, "issues being addressed include employment, workers rights, affordable housing, education, municipal services, immigrant rights and a host of others" (Head and Guerrero 1991, 11–12). Thus, poor housing, health care, and education are all considered environmental problems, which makes the environment more meaningful to those with these different concerns.

The new definition of the environment as "where we live, work and play" is important to the work of the SouthWest Organizing Project. Few issues lie outside this definition of the environment; it includes everything from crime and education to immigrant rights and police abuses.[11] Richard Moore has written that "there is much grassroots activity in the Southwest and throughout the country that addresses issues of environmental degradation in communities of color. These efforts have not received recognition, in part, *because they have not been carried out by 'environmental' organizations*" (Moore 1991a, 10).[12] The SouthWest Organizing Project emphasizes that its involvement with the environment is not a new development, but one that has been going on for decades.

The project has connected with its own history of organizing, using its previous work as a way of grounding its involvement with the environment. For years, groups "have defined environmental issues as an integral part of community, labor, or civil rights causes. The pesticide struggle has for many years been spearheaded by Latino farmworkers. Occupational health and safety issues have been addressed by labor unions, or in the case of the border region, by immigrant rights groups. Lead-based paint poisoning has been addressed by community-based housing organizations" (Head and Guerrero 1991, 9–10). For the SouthWest Organizing Project, the issues of the environment, poverty, and the housing conditions of local residents are so tied together that their new definition of the environment as "where we live, work and play" touches on virtually every aspect of the community's problems.

The leaders of the project maintain that their communities have always been involved with the environment. As have so many groups, the SouthWest Organizing Project has challenged the assumption that their communities are not interested or are indifferent to the environment. As Richard Moore reflects, "this may not be traditional environmentalism but it's the kind that has concerned people of color for centuries" (Martinez 1991a).[13]

The misleading impression that poor or working-class Latino communities in the Southwest are not involved with the environment, according to leaders of the project, is rooted in differing perceptions of what constitutes an environmental problem as well as a lack of awareness by mainstream environmental organizations of these communities' long history and involvement with the environment.[14] "People of color have historically been involved in struggles for environmental justice. The false idea that such involvement has not taken place is rooted in the narrow definition of environmental issues by environmentalists. People of color, on the other hand, have defined environmental issues as an integral part of community, labor or civil rights causes" (Head and Guerrero 1991, 9).[15]

The connection of the environment with many other concerns is a cornerstone for the SouthWest Organizing Project, which emphasizes—as does the Gulf Coast Tenants Organization and other groups in the environmental justice movement—that their communities have long been concerned with the environment. "In the 1960s, we were involved in tenants' rights organizing, and many of us are still involved in that today. There are over nine hundred thousand housing units in this country today that are painted with lead-based paint. What that means to us is that our children have become very ill throughout the years from eating the paint that chips off the wall in our housing projects. But we have never perceived those as environmental issues, we have perceived them as tenants' rights issues" (Moore 1991b, 62). "The organizing principle of the SouthWest Organizing Project," writes its founder, Richard Moore, "is to demystify what 'environment' means."[16] It is this refraction of the environment through organizing around issues of social justice, poverty, and housing that is uppermost in what Moore and other leaders of the project describe as the demystification of the environment. This is an environmentalism that is interwoven with the fabric of everyday life for groups throughout the Southwest.

In drawing from the history of previous movements in the Southeast and using this history to frame the environment, the SouthWest Organizing Project is creating what it refers to as a new definition of the environment as not only the places where people "live, work and play" but that also implies a historical understanding of those circumstances that is rooted in centuries of commercial expansion and development. The group is uniquely situated so that its organizing draws from earlier traditions of political activism as it redefines the environment as a part of the culture of the Southwest.

CONCLUSION

The work of the Gulf Coast Tenants Organization and the SouthWest Organizing Project reflects a connection of the environment with some of the most persistent problems of racism, discrimination, and poverty. This connection of the environment with many different problems (as argued in this chapter and the previous chapter) is indispensable to the work of these groups and reflects the possibility of building new environmental politics in these communities. The work of these groups also reflects the importance of tenants, civil rights, and community groups in the environmental justice movement. In both instances, groups that had already been organizing in their communities for years expanded their work to incorporate the environment.

The framing of the environment by the Gulf Coast Tenants Organization and SouthWest Organizing Project in terms of its connection with other concerns in these communities is an important example of the framing found in the environmental justice movement. In framing the disparate incidence of environmental hazards as environmental racism, the leaders of these groups have brought racial explanations to the foreground in their portrayal of the environment. The leaders of both the Gulf Coast Tenants Organization and SouthWest Organizing Project have expanded their work to include environmental issues, largely by framing them as part of a much longer history of discrimination. In the process, these groups have also framed a new understanding of environmentalism using a language of justice, racism, and rights, and have based this understanding on a new definition of the environment as "where we live, work and play."

NOTES

1. The Chicano Earth Day was dedicated to the struggle against what its participants referred to as *social pollution* in Albuquerque (*Albuquerque Journal* 1970).

2. Richard Moore, who would later help found the SouthWest Organizing Project, was the director of a health clinic in the South Valley.

3. Early in its work, the SouthWest Organizing Project also organized many of the former workers and residents living in the vicinity of Albuquerque's General Telephone and Electronics plant, which was alleged to have exposed workers and residents alike to hazardous wastes over the years (Martinez 1991a, 4).

4. "The loss of land and water resources weakens the economic and political underpinnings of local communities and the very cultural foundations of indigenous people here" (SouthWest Organizing Project 1994).

5. "State and federal regulatory agencies," explains the SouthWest Organizing Project, "ignore and sometimes aggravate environmental problems in poor communities" (SouthWest Organizing Project 1994).

6. Explains Jeanne Gauna of the SouthWest Organizing Project, "they perceive environmentalism as conservation but for us, it is the survival of our communities. In New Mexico, environmentalists are like the new colonizers. They love the culture and hate the people" (Martinez 1991a, 10).

7. "Some of the mainstream environmental organizations are being compromised by the very polluters that have been major contributors to the environmental crisis which now threaten us" (Head and Guerrero 1991, 8).

8. Explained a longtime writer and activist involved in the SouthWest Organizing Project, this activism in regard to the environment is rooted in "a history of

Chicano-Indian resistance to Anglo incursion that goes back to 1847" (Martinez 1991a, 4).

9. The group has been tracking the use of solvents in manufacturing, working with groups such as the Silicon Valley Toxics Coalition and workers in the industry.

10. As the leaders of the SouthWest Organizing Project write in their working paper, *Development, Environment and Race in New Mexico* (1994), "New Mexico is best described as a colony of the United States. Since its incorporation into the United States in the 1840s, New Mexico has been economically dependent on mineral and forest extractive industries and, since the 1940s, the United States military."

11. Wrote the SouthWest Organizing Project in its *1981–1991: A Decade of Struggle for Justice*, "we have witnessed environmental, economic and social devastation in our communities, whether from toxic poisoning caused by industry or the military, substandard housing and crumbling infrastructure, union busting, or racist brutality by those who are supposed to serve and protect us" (1991, p. 3).

12. Write Louis Head and Michael Guerrero, "the environmental movement has shown little willingness to recognize the legitimacy of struggles to alleviate the poisoning of communities of color" (Head and Guerrero 1991, 10).

13. Richard Moore, a founder of the SouthWest Organizing Project and director of the Southwest Network for Environmental and Economic Justice, has written that "to the mainstream environmental movement, we say emphatically that we have been involved in environmental issues for many, many years, environmental issues as we perceive them, as poverty issues, survival issues, and issues of racial and social justice" (Moore 1991b, 63).

14. Explains Richard Moore, "the problem is that we perceive environmental issues as one thing and they perceive them as another. Those people may not perceive [a] sewerage plant as an environmental problem but they do not live there. Most environmentalists have no contact with the workers in a plant that is polluting, or in a community where people live with the problem. We do" (Martinez 1991a, 10).

15. Explained a long-time activist involved with the SouthWest Organizing Project, "it is time to dump the myth that people of color never care about the environment because they are too concerned with mere survival. If anything, their survival is an environmental issue" (Martinez 1991a, 4).

16. "At any community meeting," concludes Moore, "just ask 'How many people here have lived near a chemical plant? Please stand up.' Then ask, 'How many have lived near a slaughterhouse?' Then, 'How many have lived near a sewerage plant?' And pretty soon, everybody in the room will be standing. We are getting past the perception of environmentalism as a middle-class, white movement that excludes the human factor" (Martinez 1991a, 10).

4

Working-Class Politics and Environmental Consciousness in the Environmental Justice Movement

The environmental justice movement is for the most part associated with the activism of Latino, African-American, Native American, and other communities. It is also distinguished in part by the struggle to form a new definition of the environment by working-class people in their unions and their neighborhoods. Working people living in the shadows of industry now routinely deal with the environment and its relation to their health and that of their families.

Approximately twenty-five million workers each year are exposed to some form of hazardous materials in the workplace. Of these, an estimated fifty to seventy thousand die from complications related to their exposure to materials on the job. And some of those who deal with these hazards in the workplace are often the first to be exposed to chemicals that later show up in sites near their homes.

FRAMING THE ENVIRONMENT IN A WORKING-CLASS CONTEXT

There is a growing awareness among scholars about the struggle of working-class communities to reclaim a stake in the fight for the environment. Careful studies of working-class culture have found a deepening regard for the environment in music, magazines, and popular culture, confounding perspectives that narrowly identify environmental concerns with more affluent portions of the populace.

The environmental justice movement has broadened beyond race and ethnicity to include the class inequalities that make some communities more vulnerable to environmental hazards than others. In some parts of the country, labor unions and their members are working with local residents against hazardous wastes or pollution in their communities despite the uncomfortable position in which this sometimes places workers.

Much like the work of the SouthWest Organizing Project and the Gulf Coast Tenants Organization, the leaders of union locals are framing a new definition of the environment which brings the hazards for working people to the foreground. Property taxes, housing values, and occupational health are only a few of the problems that are framed by these unions as environmental issues.

Although race and ethnicity resonate in the framing of the environment by the Gulf Coast Tenants Organization and SouthWest Organizing Project, it is clear that economic class is also an important factor in the incidence of environmental problems. In most areas, those who are exposed to hazardous wastes are, not surprisingly, usually residents of working-class neighborhoods. National studies, for the most part, continue to show that class is significant in the incidence of environmental problems. Nearly forty-one million Americans live within five miles of the more than eleven-hundred Superfund sites in the nation. Of this, an estimated fifty percent are working-class and lower middle-class whites, according to findings of the study, *Toxic Wastes and Race in the United States*, published in 1987 under the auspices of the United Church of Christ Commission for Racial Justice. This and the subsequent chapter will argue that workers and their unions are framing their understanding of the environment to include the different concerns that affect the communities in the vicinity of industries.

INDUSTRY AND THE STRUGGLE FOR THE ENVIRONMENT IN LOUISIANA

Amidst the expanse of Louisiana's wetlands along the Mississippi River, the members of Oil, Chemical and Atomic Workers Local 4-620 and their families led a coalition of chemical workers, local residents, and environmental groups in the longest union lockout in the nation's history. Of an array of projects that emerged from this five-year lockout, one of the most innovative was the incorporation of the environment into the organizing of the chemical workers union. This work—to incorporate the environment into the organizing of a labor union local of chemical workers in the Louisiana wetlands—reflects the framing of the environment as part of more immediate concerns of local residents, in particular their health problems, an inequitable property tax system, and the lack of an adequate municipal infrastructure in many of the towns along the Mississippi River.

The industry along the banks of the Mississippi River in Baton Rouge, with the smokestacks and gas flares that dominate the city's skyline and provide a backdrop for the skyscraper state capitol built under the administration of Governor Huey Long in the 1930s, reflects the influence of the chemical industry in the economic and political life of Louisiana. The industry shapes much of the economy in Louisiana, accounting for nearly half of the state's gross annual product. "Local nightly news programs," writes Carl Freedman, "track the daily price fluctuations of West Texas crude as faithfully as they do the stock market, the weather or the sports scenes" (Freedman 1992, 23). Accidental leaks and stories of explosions at the plants are commonplace on the nightly news, as are stories of residents being forced to evacuate their homes because of these accidents.

Ascension Parish, just south of Baton Rouge, has some of the highest releases of airborne emissions and industrial waste along the Mississippi River. A 1986 study by the Sierra Club found that nearly two hundred million pounds of chemicals are released annually from its eighteen plants. In 1992 alone, local residents documented more than three hundred accidents at the parish's eighteen chemical plants. The parish, according to the Environmental Health Network and the National Toxics Campaign Fund, also ranks as one of the highest in the nation in deaths from lung cancer (Lewis, Keating and Russell 1992). These health problems are even more disconcerting because of the lack of an adequate municipal infrastructure to assist local residents with these problems.

THE LOCKOUT AND THE OIL, CHEMICAL AND ATOMIC WORKERS

Ascension Parish, a parish on the east bank of the Mississippi River and thirty miles south of Baton Rouge, is home to several small towns where the life of residents changed with the arrival of the chemical industry. The town of Geismar is home to more than a dozen chemical plants, three hazardous waste sites, and numerous landfills as well as oil waste lagoons. There are also smaller abandoned waste sites, making it among the most environmentally threatened regions of the state. The flames of the gas flares and the constant procession of tanker trucks make this otherwise quiet, rural parish busy throughout the day, while second and third shifts run at the chemical plants through the night.

Geismar is similar to many small towns along the Mississippi River. With its access to the river and its large tracts of undeveloped land, the town is home to several of the largest plants in the chemical industry. In 1957, the German-based chemical firm BASF built a plant in Geismar, taking advantage of the state's property tax laws and its proximity to the Mississippi River.

The BASF plant is among the few plants along the Mississippi River represented by the Oil, Chemical and Atomic Workers International Union. The union local was organized in 1959 shortly after the facility was opened. In the early 1980s, following years of cooperation between the union and the management of BASF, Local 4-620 saw indications that management was moving to challenge the union's certification. In July 1983, the union's small office at the plant was closed by management.[1] In contract negotiations, management also allegedly demanded concessions from the union that included a one-year wage freeze for all employees, a reduction in health care benefits, wage reductions for plant maintenance workers, and changes in seniority and retirement benefits.

With union leaders and plant management unable to resolve their differences in negotiations, a lockout at the Geismar plant began on June 15, 1984 when the contracts of 357 members of Local 4-620 expired and they were locked out of the plant. Workers argued that management was doing this as part of a nationwide effort to decertify the union, and filed the first of a series of unsuccessful injunctions against the lockout with the National Labor Relations Board.[2] In the meantime, BASF replaced its workers at the Geismar plant with nonunionized workers, some of whom, Local 4-620 alleged, were construction workers inadequately trained to deal with the hazards of a chemical plant.

THE ENVIRONMENT IN ASCENSION PARISH, LOUISIANA

The early months of the lockout were the most difficult for the Oil, Chemical and Atomic Workers Local 4-620, which initially found few supporters either in the community or from other unions in Louisiana. The lockout began during a period of sluggishness in the oil refining industry, which meant that few unions in the state were in a position to support the workers.

From the early months of the lockout, the national leadership of the Oil, Chemical and Atomic Workers as well as the union local in Geismar framed their struggle as critical to the survival of the union. The national leadership of the Oil, Chemical and Atomic Workers International saw its declining membership across the country as a backdrop for the lockout. Membership had reached a low in the union's history, with less than twenty percent of the workforce in the chemical industry unionized in the 1980s. By the end of the decade, its membership had fallen to less than one hundred thousand members. In the early months of the lockout, union leaders alleged that BASF was among those firms in the chemical industry pursing a policy of breaking the union in its plants.

Within several months of the lockout, as Local 4-620 realized that they had been too confident in anticipating a quick settlement with management, union members turned their attention to the residents who lived in the area surrounding the BASF plant. This was the first time that the union had sought to work with nearby residents, particularly those who had been raising questions about the environmental and health consequences of the plant's operations.

Consideration of the environmental and health problems of the industry was difficult for many of the workers in Local 4-620. Early on they denied that the chemicals in the plants were connected with any significant health threats. Workers were hesitant to accept the idea that their supervisors had provided them with limited information on the health threats of the chemicals they were working with in the plant. "We felt that if there was anything that was going to hurt us, the company would tell us about it. We thought that the management would print it on the bulletin board and we would know about any problems," explained one worker. "In the early days, we had workers who would eat their lunch on bags of asbestos. We never knew this could hurt us," recalled another. As the lockout continued, workers expressed disbelief that they had been misled by management and began to investigate the health threats of the chemicals they had worked with over the years.

The Oil, Chemical and Atomic Workers Local 4-620 approached the state chapter of the Sierra Club and the newly formed Louisiana Environmental Action Network. Workers had to build the confidence of local residents and assure them that the union was committed to working with the environmental and health concerns of residents.

Workers in Geismar drew from the history of the Oil, Chemical and Atomic Workers International and its involvement in struggles for occupational health and safety. The union has had a long history of involvement in workplace health concerns since its formation in March 1955. The union's national leadership had led the right-to-know movement and worked closely with the Sierra Club. In a letter to the Louisiana Environmental Action Network in Baton Rouge, Local 4-620 president Esnard Gremillion explained, "we will be exactly where we have always been, in the center of the movement to unite workers and communities around the environmental issues that affect us all." Union leaders enclosed newspaper clippings on the history of the Oil, Chemical and Atomic Workers, emphasizing its involvement with the Sierra Club and other organizations.

This history of the Oil, Chemical and Atomic Workers International made possible the involvement of locked-out workers at BASF with the nearby community. The relationship between local residents in Geismar and workers at BASF, explained a worker at the plant, took hold "once we started teaching the environmental activists about the history of the union and convincing them that the union wasn't only concerned with stopping the lockout. The union had been a part of the environmental movement for decades." This history of the Oil, Chemical and Atomic Workers was important for its organizing, insofar as residents who knew relatively little about the union learned more about its history while workers became less suspicious of environmental groups.

As workers investigated the local residents' health concerns regarding the chemical plants, what knowledge they had of the health effects of the chemicals they worked with over the years was shared with community residents. The union published materials for residents detailing the health effects of the chemicals used in the BASF plant. Workers also took advantage of documentation that was newly available under the provisions of the 1986 Superfund Reauthorization Act.

Few workers at the BASF plant had had any previous involvement with environmental groups. As one worker explained, "we thought that environmentalists wanted to put us out of work." "I knew nothing about environmental groups until the lockout. Most of us really didn't think that the problems we faced were environmental problems," reflected another union member. Workers gradually overcame this mistrust. Many felt that the plant management at BASF had misrepresented the work of environmental groups such as the Sierra Club. Eventually, this process of building trust among workers led to changes in the way most of the workers viewed the environment.

Almost a year into the lockout, Local 4-620 began to investigate the environmental problems of the BASF plant as part of what it referred to as its corporate accountability campaign. The union brought in national organizers and sought advice from the National Toxics Campaign. In the meantime, the National Labor Relations Board rejected more than fifty injunctions filed by the union; therefore, this organizing in Geismar around the environment was in part a response to the stalled litigation and federal labor grievances of the union.

In its literature, Local 4-620 reminded community residents of the consequences of the lockout for the well-being of residents. Workers did everything they could to frame the lockout not only as a labor dispute but as a problem for the entire community. For example, workers labeled their money with the phrase "union money." As the money circulated through the community, according to one worker, "it reminded local merchants how much a part of the community the union was and made many think twice about speaking out against the union."

The Oil, Chemical and Atomic Workers held several large events, including rallies and marches at the plant itself. Locked-out workers held protests at the annual stockholders meetings of BASF and even made an attempt to purchase shares of the corporation's stock so that workers might speak at these stockholders meetings. When these attempts to influence BASF failed, Local 4-620 initiated a boycott of the corporation's products, including its audio tapes and antifreeze. The union also participated in the November 1988 march from Baton Rouge to New Orleans with the Gulf Coast Tenants Organization.

Local 4-620 distributed publications on what they claimed were releases of hazardous materials, workplace accidents, and related incidents at the BASF plant that were taking place during the lockout. Perhaps the most important aspect of this activity was the publication of the union's newsletter, *Basagram*. Its newsletter was important for the union, providing detailed information to workers, their families, and local residents. Union members published internal memorandum and other materials from BASF. The newsletter also published confidential reports from plant managers that acknowledged the declining safety conditions in the plant, casting doubt on management's assurances about the safety of the BASF facility.

The Oil, Chemical and Atomic Workers depended heavily on the families of community residents and other groups during the lockout. The hardships that workers endured were difficult for many families. Workers lost their cars and their homes, and in several instances, marriages and families were torn apart. The union spent considerable time working out payment arrangements with local banks on behalf of locked-out workers. Local 4-620 met with the families of workers each week to coordinate these efforts, often at the volunteer fire department in Prarieville.

Local 4-620 sought to prevent the company from eroding support for the lockout among workers by putting together groups that involved the families of workers. A women's support group was formed early in the lockout. Entire families often took part in the pickets at the plant. Union leaders were also resourceful in pulling together support from small businesses to help workers, distributing food donated by local merchants when workers picked up their strike benefits from the union.

A family support program was set up, with contributions from labor activists throughout the nation. The idea of a support program had been borrowed from a strike of the United Food and Commercial Workers in Austin, Minnesota. Union locals throughout the country pledged their support for workers and their families for the duration of the lockout. This provided a base of financial support and helped workers with their home mortgages and car payments.

Many workers at the plant had a long history of employment with BASF and found it difficult to question the management of the plant. As Reverend George F. Lundy and William Temmink wrote, "work at the plant is often a family affair, as sons follow fathers and younger brothers follow older brothers to work in the plant" (Lundy and Temmink 1987, 7). In the early months of the lockout, however, this connection was changed for many workers. For the first time, many saw themselves as expendable to or replaceable by a company that they had worked with for decades.

Workers and their families regularly gathered at the volunteer firehouse in Prarieville for potluck dinners. These gatherings and the support groups of the union helped workers and their families endure the lockout. They also helped the union broaden its work with the community and organize alongside local residents to question the impact of industry on the environment and their health.

FRAMING ENVIRONMENTALISM BY THE OIL, CHEMICAL AND ATOMIC WORKERS

As the lockout continued, the Oil, Chemical and Atomic Workers framed the lockout at BASF as an environmental problem. In bringing to the foreground what the union depicted as an attempt by management to reduce its labor costs through the replacement of a unionized workforce with largely unskilled workers, the union made further inroads into the community. The union argued that the use of replacement workers with no prior training jeopardized the health of those living in nearby Geismar.

Local 4-620 emphasized that the BASF plant had more accidents during the lockout. The union conducted its own research with the Sierra Club, which alleged that the BASF plant had released more pollutants than permitted under its operating permits. In March 1986, the union published another study, which found that the number of workers exposed to hazardous materials in the plant had increased over a period of less than two months since the lockout had begun. The union also documented more than eighty accidental releases of pollution in the first nine months of 1986.

Much like the Gulf Coast Tenants Organization in its work with African-American communities, Local 4-620 emphasized a connection with the history of the plantations in its framing of the environment. Since the 1920s, the chemical industry had constructed many of its facilities on the landholdings of former plantations. Some of the largest chemical manufacturers still have these historic mansions inside the fence lines of their plants. In its publications, the union made every effort to connect its struggle with the history of the towns along the Mississippi River. As Richard Miller, a national organizer for the Oil, Chemical and Atomic Workers, later reflected, "what's happened in Louisiana is that the chemical companies have taken the place of the old sugar cane plantation owners and are now the new masters. The two-class system still prevails with local blacks firmly mired at the bottom. In return for extracting what they need, the companies pay off the state and parish officials and placate the locals with a few crumbs and an occasional crawfish boil. Meanwhile, the infrastructure of the towns continues to disintegrate, the people sicken and die, and the air and water become more toxic" (Beasley 1990). The union's publications likened the experience of workers in the chemical industry to that of the working conditions on the plantations.

For union leaders, the relationship of the plantations to the chemical industry was important in their framing of the environment. For many of the residents in the town of Geismar as well as for African-American workers in Local 4-620, the chemical industry is seen as a reminder of the historic inequalities in the region. "The economic relationship is very much the same as the plantations. The communities that always lived alongside the plantations now live with the plants," reflected an African-American member of the union local.

Local 4-620 emphasized the likelihood of a major accident in the chemical industry in Louisiana, particularly in the wake of the BASF lockout. With mounting concerns about an accident at the plant, the Oil, Chemical and Atomic Workers framed the chemical industry as connected with the tragedy in Bophal, India. This included placing "Bophal on the Bayou? Stop BASF Before They Stop You" billboards along the heavily traveled highways that intersperse the wetlands throughout this part of Louisiana.

The union also used nationalist themes in its work with the environment. In their literature, the union used the foreign ownership of BASF as a way of galvanizing workers. Billboards in Ascension Parish included the prominent theme, "Stop the Foreign Oppression of Louisiana Workers." Workers contrasted the world-wide stature of the German-owned chemical firm with their own organization, framing the corporation as a "huge multinational corporation" and the chemical workers union as an "underdog." And much as the Gulf Coast Tenants Organization and other groups in Louisiana had done, Local 4-620 described Louisiana as a "resource colony" or even a "Third World country" in its publications.

The leaders of Local 4-620 also went out of their way to frame the problems of the industry as integral to the health problems of the town of Geismar. Working with local residents, union activists documented the health problems of the nearby community. Locked-out workers also conducted their own research on the industry and monitored its releases. Members of the union monitored chemical releases in the evenings and late at night, when such releases were common and when local residents complained the most about their health problems. With respirator masks and air monitoring equipment, workers spent long evenings in the wooded areas outside of the plants documenting the release of pollutants. In the towns along the Mississippi River, locked-out workers also held meetings with local residents in school auditoriums, churches, and volunteer firehouses to address the environmental and health consequences of the industry.

PROPERTY TAXES AND THE ENVIRONMENT

The environment was cast by the Oil, Chemical and Atomic Workers in terms that connected it with the economic livelihood of residents in the town of Geismar, particularly their frustration with the lack of adequate community services. The lack of emergency services, inadequate zoning, and housing located close to the plants were only a few of the problems framed by the union as part of the larger problems posed by the plants. Even with the employment created by the chemical plants in Louisiana, the industry and its impact on the economy of the state was framed by the union as offset by property tax exemptions that reduced the state's revenues. The union therefore focused on the property tax exemptions in Louisiana and their impact on schools and emergency services.

The industrial tax exemptions were a significant part of the framing of the environment by Local 4-620. Formed in the 1930s, the property tax exemptions weakened the Louisiana public school system, its roads, emergency services, and law enforcement. Many of the towns along the

Mississippi River lack a significant infrastructure even when surrounded by some of the most prosperous corporations in the world. From 1980 through 1989, corporations in Ascension Parish were alleged to have received more than ninety million dollars in tax exemptions.[3] This led union members to form the Louisiana Coalition for Tax Justice, which shares office space with Local 4-620.

The Louisiana Coalition for Tax Justice grew out of the Oil, Chemical and Atomic Workers union. As part of its organizing, the union studied the property tax exemptions. Workers found that BASF had received a property tax exemption for a plant that had been closed years before by BASF management. Members of Local 4-620 were able to force BASF to pay more than $225,000 in back taxes on a herbicide plant it had closed several years earlier (Leonard and Nauth 1990, 44). From this work, the group launched a statewide campaign to reduce tax exemptions. Esnard Gremillion, Sr., the president of Oil, Chemical and Atomic Workers Local 4-620, became the first president of the Louisiana Coalition for Tax Justice.

In its 1992 study, *The Great Louisiana Tax Giveaway: A Decade of Corporate Welfare, 1980–1989*, the Louisiana Coalition for Tax Justice offered the first comprehensive study of the state's industrial property tax exemptions for what were described as "Louisiana's corporate tax dodgers." The "Fortunate 50," as the chemical plants were described, had been given hundreds of millions of tax exemptions under a five-year exemption program that was almost always renewed for an additional five years by the state. In all, the state was alleged to have lost $1.28 billion in tax exemptions to the chemical industry alone (Stone and Gibbs 1992, 3). Tax exemptions granted to six of the state's largest polluters were estimated to have cost the taxpayers of Louisiana in excess of $500,000 for each job these companies had created.

Local residents in Ascension Parish took advantage of the Local 4-620 workers' involvement with their community to politicize these concerns in a way they had not pursued before. Members of the union helped form the Ascension Parish Residents Against Toxic Pollution. The union also worked to set up a water quality board in Geismar, making Ascension Parish one of the few places in the nation where union activists were deeply involved with local residents in working to develop a municipal infrastructure.

The town of Geismar is like so many of the unincorporated areas in the parishes along the Mississippi River. Many residents are elderly and have had little previous connection with the chemical workers union, yet through the determination and the outreach of both union members and a handful of residents, groups were organized in the community to work on environmental problems. Local 4-620 has also sought to frame the environment in Ascension Parish as an economic problem, due to both impoverishment as well as the lack of an adequate infrastructure to monitor contaminants or deal with the health problems of local residents. Their lack of adequate services and infrastructure, in turn, is more often than not reinforced by the fact that the environmental pollution by industry diminishes property values, thereby further straining the revenues for local services.

Branching out from this organizational work, the Oil, Chemical and Atomic Workers has broadened its scope to include addressing the accountability of industry to workers and local residents in Geismar. The Louisiana Coalition for Tax Justice and its work against the property tax exemptions has strengthened the hand of Local 4-620 in communities that may be more concerned with inadequate roads or underfunded schools than with the environment.

The Oil, Chemical and Atomic Workers and their work with the environment, workers later alleged, led to pressure on the management of BASF to resolve the dispute. As a worker in the union reflected, "environmental issues made them come to the table. We framed the pollution problems from our plant in a way that threatened the whole industry in Louisiana." Several plant managers in Ascension Parish were reportedly concerned that the lockout would spark similar organizing in other chemical plants.[4] Plant managers at other chemical plants, union leaders later learned, distanced themselves from the BASF plant and its management. Local 4-620 leaders emphasized that the petrochemical industry was sensitive to the publicity associated with the union. As a worker in the lockout later reflected, "we turned industries against industries." Another worker reflected that BASF wanted to settle the lockout to "put us back to work so we would stop testing for air pollution and putting up billboards along the highway. They were afraid we would keep exposing the dangers of the industry."

Through the support of Local 4-620 that arose across the country, hundreds of thousands of dollars were made available to workers and their families. The Oil, Chemical and Atomic Workers' national leadership spent an estimated three and a half million dollars on the lockout (Nauth 1992). The workers in Geismar also had the support of union locals across the country. And an activist involved with the lockout speculated that former Louisiana state representative David Duke's campaign for governor had been useful for bringing in financial assistance from outside of the state.

For the most part, the leaders of the Oil, Chemical and Atomic Workers Local 4-620 were employees with twenty years or more seniority at the company, who brought decades of experience to the lockout. Workers who founded the union in the 1950s and those whose families had a lifelong stake in the union, in the words of one, "brought an integrity to the fight. Whatever support we got from outside, we had to win this on our own." Workers brought years of experience to the lockout, which built their sense of solidarity.

The perseverance of union members in the five-year lockout was even more remarkable given that Louisiana had one of the highest unemployment rates in the nation throughout the 1980s. With the recession in the refining industry in Louisiana, many anticipated that the workers would quickly reach a settlement with BASF. In the end, though, the company underestimated the resolve of workers. The management of BASF also underestimated the potential for the families and for community groups in Geismar to come together around the union, particularly as it became involved in property tax, municipal services, schools, and the environment.

THE POLITICS OF WORKING-CLASS ENVIRONMENTALISM IN GEISMAR

The environmental problems in Louisiana have been framed by Local 4-620 as inherent parts of the inadequate conditions in the workplace and in the community. Addressing property taxes is only one of the ways that the union frames its work with the environment. The work of the Oil, Chemical and Atomic Workers is to recast the environment in terms that connect it with the many problems in the towns along the Mississippi River. The framing of the environmental problems as part of the accountability of the chemical plants to communities, as well as the connection of the industry with the history of the region along the Mississippi River, was central to the work of the Oil, Chemical and Atomic Workers.

Local 4-620 is only one of several union locals in the chemical plants along the Mississippi River. For this reason, it had to rely on support from the families of workers and other groups. The determination of workers, their families, and the union was "cemented by the strong community bonds in the Delta region and Louisiana's populist political heritage" (Nauth 1992).[5] Cementing it also were the lifelong friendships, the loyalties, celebrations, and even the tragedies that had confronted the chemical workers as individuals and as families.[6]

The union framed its organizing to connect the environment with the economic and the health concerns of local residents. By framing labor grievances as an environmental problem, Local 4-620 was also able to politicize the labor dispute in a way that drew local residents into the organizing of the union. The work of the union, moreover, did not end after the dispute was settled. The Louisiana Coalition for Tax Justice is an example of the union's continuing effort to make connections between the environment and community and workplace concerns.

The collaboration between the union and environmental activists was not evident at the outset of the lockout, yet the projects started by the Oil, Chemical and Atomic Workers in the surrounding communities endured even after the settlement. Workers in the union have voted each year since the lockout to use a portion of their union dues to monitor the pollution in Ascension Parish by the chemical industry. This research, conducted under the auspices of the Louisiana Labor/Neighbor Alliance, is jointly supervised by the union and the staff of the former National Toxics Campaign. Local 4-620, working in conjunction with Ascension Parish Residents Against Toxic Pollutants, is also active in efforts to establish pollution-monitoring systems.

From years of organizing in Ascension Parish, the Oil, Chemical and Atomic Workers had built a trust among local residents that, in the organizers' view, allowed the union to work with the environment where few other groups have been able to do so. "Now we can go into communities and talk about environmentalism where the environmental groups could never go. So many of these communities still do not trust environmental groups, but as a union we can work with this issue in the communities because they know what we've been through," explained one worker after the lockout.

During the successful effort of the union to end the longest lockout in the history of the United States, the Oil, Chemical and Atomic Workers framed environmental issues in a way that rendered it part of the concerns of the nearby community. The most significant aspects of this were the ways in which the local union framed its work as part of its history of involvement with occupational health as well as what it framed as the connections of the chemical industry with the health concerns, the industrial property tax exemptions, and the histories of the towns along the Mississippi River.

NOTES

1. As the Reverend George F. Lundy, S.J. and William Temmink explained in a later account, "on July 19, 1983, [BASF Plant Industrial Relations Manager Francis Richard] Donaldson told [OCAW Local 4-620 President Esnard] Gremillion that things were going to be different. The union office at the plant would be closed. The company would no longer allow Gremillion to represent workers while he was still on the clock. When Gremillion asked for time off to conduct union business, he was told that he would have to make his request in writing at least five days before the actual time off requested. In short, there would no longer be union representation at Geismar" (Lundy and Temmink 1987, 3).

2. "The Reagan era's pro-business National Labor Relations Board, backed by the federal courts, consistently ruled against unions and eventually rejected all 50 charges filed by the Oil, Chemical and Atomic Workers during the lockout" (Leonard and Nauth 1990, 37).

3. In Ascension Parish, the union's study found that the building and land values of industrial plants in the parish paying no taxes was a "mind boggling" $1,274,402,415. The union estimated that this cost Ascension Parish residents and taxpayers $96,622,771 between 1980 and 1989, and $9,909,496 in 1989 alone. All of these funds could have been used for infrastructure improvements such as much-needed sewer and water systems in the parish (Gozan 1993, 11).

4. According to Richard Leonard, the Special Projects Director of the Oil, Chemical and Atomic Workers International Union, and Zach Nauth, the former executive director of the Louisiana Coalition for Tax Justice, "toward the end, the pressure was coming not only from unions and citizens but from other companies. Many Louisiana companies felt the German giant brought down trouble on the en-

tire petrochemical industry with its stubborn lockout. One locked-out worker's wife, who is employed by a nearby chemical company, confided that her bosses were blaming stricter hazardous waste regulations on BASF. No doubt the chemical industry hopes the settlement will deflate the growing pressure for corporate and government accountability throughout Louisiana" (Leonard and Nauth 1990, 46).

5. Explained one leader of Local 4-620 in the wake of the lockout, "we are working towards a long range goal that is nothing short of the realignment of power in Southern Louisiana" (Gozan 1993, 11). As another explained, the union is "battling the skewed power structure that has served to keep poor and working people disenfranchised" (Gozan 1993, 11).

6. The leaders of the lockout were able to see their struggle as bound up with the health and safety not only of nearby residents but of their own families, their own children and their grandchildren. Explained one leader of Local 4-620, "brothers in other [union] locals are a lot more reluctant to holler at the environmentalists than they used to be. We have helped workers understand that holding your employer accountable for pollution is not only good public relations in a corporate campaign but is necessary for the health and futures of our children and grandchildren" (Gozan 1993, 10).

5

The Labor/Community Strategy Center

The Labor/Community Strategy Center emerged from the organizing of United Auto Workers Local 645 in 1982 to keep open a General Motors plant outside of Los Angeles. With more than a decade of experience in organizing, the Labor/Community Strategy Center is now expanding its efforts to include the environment as an integral part of the many problems of working people in Los Angeles, framing it beside more recognizable problems such as unemployment, traffic congestion, waste incineration, and inadequate mass transportation.

ENVIRONMENT, INDUSTRY AND THE DEMOGRAPHICS OF POLARIZATION IN LOS ANGELES

Postwar immigration and industrialization in Los Angeles resulted in one of the most prosperous as well as the most diverse cities in the nation. The recent influx of immigrants from Asia, Latin America, and elsewhere is crucial to the landscape of the city. The hills of the city are home to some of the nation's wealthiest suburbs, while many more live in some of the poorest, most violent neighborhoods in the nation. Former gang member "Monster" Kody Scott's account of growing up in Los Angeles depicts a city locked in a state of near-war with its youth, whose gangs are as much a consequence as a cause of the crumbling social fabric in many of its neighborhoods. Amidst this, it is unlikely that the environment would be an immediate concern, yet the work of a handful of groups—including the Labor/Community Strategy Center—is leading to a new understanding of the environment that frames it as part of the lives of working people in Los Angeles. The environmental and health problems in Los Angeles are some of the worst in the nation. Its densely populated neighborhoods and its suburbs have led to some of the nation's worst air pollution, while the hills surrounding the city prevent many of the pollutants from dispersing. As a result, the city has some of the nation's highest ozone concentrations.

The Los Angeles metropolitan area has some of the worst air emissions from industrial and automobile pollutants in the nation, yet lost in much of the attention to this issue is their impact on the poor. Air pollution in many of the lower income neighborhoods of Los Angeles is linked to an influx of smaller industries. The past decade has seen an explosion of small manufacturing firms in the city's once-thriving industrial neighborhoods. Industries that left this region in the early 1980s are being replaced by small chemical plants, electronics outfits, furniture factories, and warehouses.[1] These firms are taking advantage of the influx of immigrants from Latin America, Asia and elsewhere.[2] The Labor/Community Strategy Center depicts the workers in this low-wage economy in Los Angeles as being on the front lines of what it refers to as a "new environmentalism." It is in this context that the Labor/Community Strategy Center is working on environmental problems in communities where the environment is often overshadowed by other concerns.

FROM PLANT CLOSING TO ENVIRONMENTALISM IN VAN NUYS

The founders of the Labor/Community Strategy Center began their work during a plant-closing campaign in Van Nuys that lasted nearly a decade. The campaign was initiated by leaders of United Auto Workers Local 645 in 1982. Five of California's six largest automobile plants had been closed since 1980, despite the fact that Los Angeles is one of the nation's largest automobile markets. Pete Beltran, a worker at the General Motors plant in Van Nuys, ran for the presidency of the United Auto Workers as an opponent of the plant closing and was elected as the first Latino president of Local 645.

In the wake of his election, Beltran and a group of workers at the plant began to organize against what they saw as its impending closure. United Auto Workers Local 645 formed a Community Action Program. More than five thousand General Motors workers and their families would be left unemployed by the plant closing, and many smaller businesses, local retailers, and suppliers would also face its consequences. The Community Action Program was likened to a union-in-exile, as workers held meetings at local restaurants to reach out to the community and organize against the closing.

Early on, the union framed the plant closing by publicizing the impact it would have on the nearby community.[3] Because of a large African-American and Latino workforce, its organizing against the plant closure "synthesized" the labor movement with the civil rights movement. The organizers of the campaign to keep the General Motors plant open in 1982 recognized early in their organizing that the plant closing had to be framed as a community concern. They understood that it "had to be conceptualized as a community issue rather than narrowly as a workplace issue" (Mann 1990c). The union "took an issue of a possible plant closing and made it into a debate on civil rights, a community issue and a debate on worker and community rights" (Labor/Community Strategy Center 1990). The diversity of the General Motors workforce was crucial to the efforts to extend the organizing of the union into the surrounding community.

The campaign met with modest success. In November 1986, General Motors announced the closure of eleven plants in Southern California. Eventually, the General Motors plant in Van Nuys was closed in August 1992 after nearly a decade of organizing by the United Auto Workers. Despite this, organizers were confident that their campaign had made a difference. The legacy of the campaign was a group of organizers who remained involved in building on the experiences of their work in Van Nuys.

With the closing of the General Motors plant in Van Nuys, there was a sense that activists needed an issue that would tie together many of the problems raised in the plant-closing campaign. A growing number of union leaders recognized that the environment was just such an issue. Leading figures in the United Auto Workers argued that the plant closing "has broad application to a community-based toxics campaign" (Labor/Community Strategy Center 1990). On the basis of this issue, the Labor/Community Strategy Center was formed from the organizing of the United Auto Workers. The group was as committed as its name implied to building on the connections between the workplace and the community.

L.A.'S LETHAL AIR: RACISM, POVERTY AND THE FRAMING OF ENVIRONMENTALISM IN LOS ANGELES

The initial work of the Labor/Community Strategy Center was concerned with the environment in African-American and Latino communities, particularly in South Central and East Los Angeles. Early in its work, the center felt that labor unions would support work on the environment if it were raised as a concern by the residents of these communities themselves. "Industrial workers are far more likely to listen to the voices of the mothers and fathers who are living right across the street from a factory and refinery than they are [to] the scolding of white-collar environmentalists" (Mann 1992). On this basis, the Labor/Community Strategy Center began a door-to-door canvass to talk with residents about their concerns. The work of the center culminated in the late 1980s with a city-wide meeting of labor unionists and the publication of its first study of the Los Angeles environment. The center's organizers had conducted their own study of the air emissions in Los Angeles as part of a wider campaign to bring environmental problems into the labor movement.

A second, more extensive 1991 study by the Labor/Community Strategy Center, *L.A.'s Lethal Air: New Strategies for Policy, Organizing and Action*, looked at the impact of the environment across different groups. Ozone levels in Los Angeles were documented as being highest in working-class neighborhoods. "While air pollution impacts all people in Los Angeles, it is workers in factories, low-income people, and residents of communities of color who are disproportionately impacted by the lethal air we breathe" (Mann and the Labor/Community Watchdog Organizing Committee 1991).[4] More than half the population of Los Angeles was found to be at serious risk for disease associated with air pollution.

The publication of *L.A.'s Lethal Air* was seen as the centerpiece of organizing by the Labor/Community Strategy Center. The study included pictures of the residents of Los Angeles' poorest communities, particularly children and others vulnerable to air pollution such as the elderly and people with immune system deficiencies. The Labor/Community Strategy Center concluded with the idea of what it called a "corporate accountability campaign."

FRAMING ENVIRONMENTALISM IN THE LABOR/COMMUNITY STRATEGY CENTER

Much like the Gulf Coast Tenants Organization and the SouthWest Organizing Project, the Labor/Community Strategy Center is framing the environment as an integral part of the workplace concerns in Los Angeles as well as the accountability of industry to residents in the city. "The main obstacle to resolving the environmental crisis," according to the Labor/Community Strategy Center, "is more political than chemical" (Mann and the Labor/Community Watchdog Organizing Committee 1991, 72). "If we want to change our chemical environment, we have to change our political and economic environment as well" (Mann and the Labor/Community Watchdog Organizing Committee 1991). In its organizing, the group frames the environment as an integral part of Los Angeles's many problems.

The work of the Labor/Community Strategy Center and its effort to frame an environmentalism that incorporates the concerns of workers and communities is part of a "highly conscious organizing model" that has connections with the earlier work of the plant-closing movement. The environment is framed in such a way that it brings the experiences of poor and working-class communities to the foreground. "The environmental crisis impacts us all but not equally," emphasizes the Labor/Community Strategy Center. "Workers exposed to toxics on their jobs, people of color in those workplaces and in communities marked by industrial emissions and high-density traffic, and women in dangerous workplaces and vulnerable communities will pay the highest social cost. It is workers, women, and people of color who must generate new strategies and organizations for the environmental movement" (Mann and the Labor/Community Watchdog Organizing Committee 1991, 27). The leaders of the center have worked to frame the environment in a language that has the most relevance for these communities.

The center emphasizes the importance of connecting the environment with problems ranging from the living conditions of migrant workers to the health problems of the elderly and people with immune system deficiencies. Environmental problems are framed as "life and death" concerns. While the residents of more affluent communities can avoid environmental problems, "it is [in] working-class and minority neighborhoods [that] homes, factories, and traffic congested streets become toxic neighbors" (Mann and the Labor/Community Watchdog Organizing Committee 1991, 30).[5] The environment is framed in a language that connects it with the lives of many residents in the city. In its framing of the smog in Los Angeles, the language of class inequality is as crucial as the language of air pollution. "The air toxins are trapped under an inversion layer of warm, stagnant air while the majority of people are trapped under an inversion layer of class, race and gender exploitation" (Mann and the Labor/Community Watchdog Organizing Committee 1991, 9). The Labor/Community Strategy Center has also framed this as a part of what it calls a "demographics of polarization."[6]

In its 1991 study of air pollution in Los Angeles, the center focused on the intersections of race, ethnicity, and class in what it referred to as the "unspoken categories" of environmental health. The group also frames the environment as part of "the intensified impact of air pollution on people of color both because of the structural racism of our society in which risks are consciously distributed disproportionately and because people of color reside overwhelmingly in working-class communities. It is into these communities that our society exports many of the environmental as well as the social problems that it cannot solve" (Labor/Community Strategy Center 1993b, 11). The group has framed an agenda of what it refers to as corporate accountability of which environmental problems are a part.

The Labor/Community Strategy Center has encountered a mistrust of the postwar environmental movement in many working-class neighborhoods.[7] The center frames the environmental issues as problems of "public transportation, non-polluting economic development, housing, taxes, land use, zoning, workplace health and safety, and industrial pollution reduction" (Mann 1990b). Without this framing, the center's work with the environment could "threaten rather than motivate people."

Even with its emphasis on the rebuilding of industry in Los Angeles, the Labor/Community Strategy Center remains focused on the environmental problems of industry as well as on the creation of more sustainable forms of investment. "What is the sense of having a factory that produces zero-emission public transit vehicles or high-mileage cars if the paints, epoxies, and solvents kill the workers and surrounding residents?" (Labor/Community Strategy Center 1993a, 14). In addition to this recognition of the threats to the health and well-being of workers, the center links environmental concerns to economic development and the rebuilding of communities.

The Labor/Community Strategy Center's foremost concern is to frame the environment in a way that connects it with the interests of workers and communities. Just as the United Auto Workers transformed the plant closing in Van Nuys from a labor dispute into a community issue that focused on corporate accountability, the Labor/Community Strategy Center has also sought to broaden its framing of the environment in order to connect it with workers and local residents in neighborhoods across Los Angeles.

ENVIRONMENTALISM AND INDUSTRY IN A WORKING-CLASS COMMUNITY: THE CORPORATE CAMPAIGN IN WILMINGTON

From the outset of its organizing, the Labor/Community Strategy Center has been involved in framing the similarity of interests between workers and community residents. Its recent campaign in the East Los Angeles neighborhood of Wilmington reflects of the extent to which the group is working to build a compatibility of interests between different groups.

Wilmington is a landscape of warehouses, parking lots, and floodlit yards with cargo containers and chemical plants. For mile after mile, it is an unbroken expanse of refineries and warehouses interspersed with residential communities. The Labor/Community Strategy Center has been involved in a campaign in Wilmington as a result of an October 1992 explosion at a chemical plant there. The environmental problems that resulted are being framed as problems of workplace safety and accountability to local residents rather than simply as environmental problems.[8]

In October 1992, a hydrogen processing facility owned by Texaco Corporation exploded at a refinery in Wilmington. More than a dozen workers were injured in the explosion and an estimated two thousand health-related claims for nausea, skin rashes, and broken bones were claimed to have been reported by local residents. According to the Labor/Community Strategy Center, residents were provided with little information about the explosion. This was the kind of campaign organizers with the Labor/Community Strategy Center had anticipated in their earlier study of the air pollution in Los Angeles.

The accident in Wilmington is framed by the Labor/Community Strategy Center as an example of the accountability of industry to the local community. The centerpiece of the campaign is the center's demand for a health clinic "that specializes in the diagnosis and treatment of diseases and symptoms associated with chemical exposure" (Labor/Community Strategy Center 1994a).[9] The group also demanded an emergency evacuation plan.[10] In addition, the center worked on persuading local businesses to comply with the Community Right to Know Act. Finally, the center expressed strong reservations to what it deemed the insensitivity of local officials to nearby residents in Wilmington.[11] By framing the environment in this way, the center seeks to avoid dividing workers and the community.

Along with the local community residents and workers in the refineries concerned with the health consequences of industry in Wilmington, the center also made it a point to work alongside the Wilmington Homeowners Association on the issues of pollutants from the oil refineries, chemical plants, and the solid waste treatment facility in Wilmington. The Wilmington Homeowners Association worked with parents, students, teachers and administrators in the local schools. The Labor/Community Strategy Center helped to frame the issue of the environment as one relating to the health and well-being of schoolchildren as well as adult local residents.

In its literature on Wilmington and its leaflets announcing a boycott against Texaco, the center framed its boycott with the demand that "communities and workers have a right to decide how polluting industries will behave in their neighborhoods" (Labor/Community Strategy Center 1994b).[12] Although it demands more accountability from businesses, the group is clearly not focused on the demand for a shutdown of industry in Wilmington.

In its framing of the health problems related to industry in Wilmington, the center has made an effort to include workers. In its publications, the group describes itself as "a prounion, proworker, multiracial environmental justice organization." The group distributed *An Open Letter to the Workers at Texaco's Wilmington Refinery from the Labor Community Watchdog* (1994a) to workers and residents in Wilmington, with assurances to workers that the group does not want the industry to close down. "We again wanted to take this opportunity to let you know that our desire is to clean Texaco up and keep them here. However we believe that a dramatic change between Texaco and the community is needed in order to protect ourselves from Texaco's production method." Nowhere in the letter does the Labor/Community Strategy Center mention the environment. Instead, the problems are framed in terms of the accountability of industry to the community.

The organizing experience of the Labor/Community Strategy Center in Wilmington reflects the importance of framing in a political movement. In the literature of the Labor/Community Strategy Center, the environmental and health problems related to industry are framed in such a way as to build a sense of shared interest between workers and local residents.

INDUSTRIAL DEMOCRACY AND THE FRAMING OF ENVIRONMENTALISM

The Labor/Community Strategy Center is situating its work with the environment in the history of the union movement, especially the history of such unions as the United Farm Workers. This history is at the heart of the center's work; it is used to build an environmentalism distinct from the postwar environmental movement.

For many of the organizers with the Labor/Community Strategy Center, the histories of the labor and civil rights movements have been crucial for framing the environment. The center's 1991 study, *L.A.'s Lethal Air: New Strategies for Policy, Organizing and Action*, includes photographs of the United Auto Workers sitdown strikes in Flint, Michigan during the winter of 1937 as well as photographs of Caesar Chavez in early United Farm Workers marches.[13] It also features photographs of Rosa Parks, the Reverend Martin Luther King, Jr., and others involved in the civil rights movement in the South. These references draw together various elements from the labor and civil rights movements as a way of making parallels between these movements and more recent efforts to bring together the labor movement in Los Angeles with the environmental problems created by industry. "During the height of the labor movement in the 1930s, it was the courageous and creative actions of auto workers, such as the sit-downs in GM's Flint, Michigan plants, that sparked the growth of industrial unionism and the C.I.O. During the civil rights movement of the 1950s and 1960s, it was highly visible direct action campaigns such as the Montgomery bus boycott, the Freedom Rides, the lunch counter sit-ins, and the March on Washington that created the social climate and organized muscle to win civil rights legislation" (Mann and the Labor/Community Watchdog Organizing Committee 1991, 54). This history is reiterated in the work of the Labor/Community Strategy Center.

The center's demand that industry be accountable to workers and communities is grounded in what it calls industrial or direct democracy.[14] "During the 1930s in this country, there was a powerful labor movement that demanded industrial democracy, the idea that workers and communities could have some say in the economic life of our society" (Mann and the Labor/Community Watchdog Organizing Committee 1991, 48). The work of the group to widen the involvement of residents and workers with the environment is a part of what the organizers of the Labor/Community Strategy Center see as an extension of this tradition of industrial democracy.

The group has framed its work with the environment as part of its efforts to revive the labor movement in Los Angeles. The involvement of African-American, Latino, and Asian Pacific workers with the environment is framed as part of "labor's strategic salvation" (Mann and the Labor/Community Watchdog Organizing Committee 1991). These workers are depicted as not only shouldering the brunt of environmental problems in Los Angeles but also as being capable of "radically transforming" its labor movement.

The center avoids placing undue emphasis on the environment. It is also reassuring communities that it will not ignore the livelihoods of workers nor the communities that surround industries. "In a highly insecure job market, where, often, destitution and homelessness are only months away from a layoff slip, we understand that demands to shut down a corporation, even temporarily, will threaten the workers livelihoods, unless *additional* demands are made" (Mann and the Labor/Community Watchdog Organizing Committee 1991, 66).[15] The group thus frames the environment in a way that addresses the concerns of workers, something that the postwar environmental movement has too infrequently done, according to the center. "If the environmental movement does not put forth demands for income maintenance and serious job retraining for displaced workers, then business interests and conservative union officials will manipulate a frightened workforce to once again accept the stereotype that 'those damn environmentalists' [represent] a politics of privilege and insensitivity" (Labor/Community Strategy Center 1990, 6).[16] This way of framing the environment is crucial for the group in its organizing.

The center seeks to assure communities that it is not going to organize for the closing of industry without consideration of the well-being of workers. This is evident in the work of the group with dry cleaners in Los Angeles. Most of the dry cleaning firms in Los Angeles use hazardous chemicals such as perchloroethylene. Many also have unsafe workplace conditions. Despite the health threats to workers, the center "would *not* be involved in, for example, a march to close down a neighborhood dry cleaner on environmental grounds. Rather, we might in the future seek to involve both the dry cleaning owners and other community residents as allies in an environmental coalition to demand that chemical companies discontinue the use of perchloroethylene and produce a safe alternative" (Mann and the Labor/Community Watchdog Organizing Committee 1991, 60). By framing the health hazards posed by dry cleaning businesses as a problem of the chemicals that are produced and sold by larger companies and chemical suppliers, the center seeks to avoid dividing small businesses, local residents, and neighborhood groups concerned with the environment.

The nation's largest environmental organizations, such as the Sierra Club, are seen as out of touch with the realities of places like South Central Los Angeles, where the environment tends to be refracted through impoverishment. Environmentalists, according to the center, have a "negative history" and a "politics of privilege and insensitivity" that is too often disconnected from poorer communities (Labor/Community Strategy Center 1990).[17] The postwar environmental movement has been depicted by the group as requiring an overhaul to continue as a political force in Los Angeles.

The Labor/Community Strategy Center, along with the Service Employees International Union, has challenged what it describes as "car pooling paid for by working people," calling it a limited solution to the transportation problems in Los Angeles. As part of this challenge, the group is organizing against increases in parking costs for municipal employees in downtown Los Angeles, calling it a kind of environmentalism that is "punishing those who can least afford to pay" (Mann 1990c).[18] The group has worked with municipal employees unions to counter what it refers to as "class-biased" solutions to the traffic congestion in Los Angeles. As increases in parking fees for municipal workers have become a flashpoint for labor organizing in public sector unions in downtown Los Angeles, the center has been on the front lines, using this issue to form a more inclusive understanding of environmentalism.

Much like its criticism of public transportation for being inattentive to the concerns of working people and the poor, the group has also challenged the limitations of recycling programs. "We want to change the debate from individually purchasing environmentally safe products to collectively demanding a ban on those products that harm the environment" (Mann and the Labor/Community Watchdog Organizing Committee 1991, 71). With its criticism of mass transportation and recycling, the Labor/Community Strategy Center introduces a class-based perspective to its analysis of mass transportation, recycling, and the environment.

The center has looked at mass transportation not only as a way of addressing the environmental problems in Los Angeles but as a way of rebuilding the manufacturing sector of the city. The group is linking transportation to unemployment in Los Angeles.[19] It advocates "environmentally-sound production of technologies, focusing on solar electricity, non-polluting, prefabricated housing materials, electric car components, and public transportation vehicles, both buses and trains" (Labor/Community Strategy Center 1993a).[20] Even in its proposals for economic revitalization in communities such as South Los Angeles, the Labor/Community Strategy Center is bringing the environment, public transportation and impoverishment together.

ENVIRONMENTALISM AS "STREET LEVEL POLICY": REDEFINING THE ENVIRONMENT IN THE LABOR/COMMUNITY STRATEGY CENTER

The work of the Labor/Community Strategy Center is not only concerned with framing the environment in Los Angeles as an integral part of impoverishment and inequality but it is also trying to redefine what the environment itself means. As an activist with the group reflected, "much of this has to do with the way in which issues are presented to people. If we only talk about the environment, we are never going to involve people in places like South Central. We have to find a way of talking about the environment that includes these communities." This is done by connecting the environment with the poverty and the living conditions in Los Angeles.

From its years of organizing, the Labor/Community Strategy Center is attuned to the importance of framing the environment in such a way as to incorporate many problems in these communities. It is, in short, an environmentalism that plays itself out as much in the neighborhoods of Wilmington as it does in the municipal employee parking lots of downtown Los Angeles, where activists have pressed the city's leadership for more farsighted approaches to the transportation and air pollution problems of the city.

The leaders of the center have learned to frame the environment by defining it as "where we live, work and play." Perhaps the most succinct expression of this is in the conclusion of the Labor/Community Strategy Center's *L.A.'s Lethal Air: New Strategies for Policy, Organizing and Action*. "We no longer have the luxury of defining ourselves in isolation as labor, community, civil rights, women's, and environmental movements. We have a symbiotic relationship. We will either learn to work together in this society or we will die apart" (Mann and the Labor/Community Watchdog Organizing Committee 1991, 72). This reflects the group's broadening of the definition of "environment" to encompass a host of issues of more immediate concern to labor unions and community residents.

The center frames the leadership of workers and people of color in what the group refers to as the "new environmentalism," an environmentalism that involves workers, neighborhood associations, and others. "Factory and office workers, high school and college students, women, Latinos, African-Americans, Asian-Americans, Native Americans, white working people, farm workers working with pesticides in the fields, and inner-city residents facing air pollution, waste incineration, and groundwater contamination must become the leaders of the new environmentalism" (Mann and the Labor/Community Watchdog Organizing Committee 1991, 58).[21] Not only must the labor movement be remade to include the environment, but the environmental movement must broaden itself to include a criticism of the lack of accountability of businesses to local communities.

From the perspective of organizers with the center, the environment is seen as a "point of entry" for rebuilding the labor movement in Los Angeles into a movement capable of confronting the economic forces reshaping the city. The environment is not elevated above other issues but framed alongside these other concerns.[22] In this way, the Labor/Community Strategy Center advocates the need for a coalition of labor unions and activists that will bring environmental issues to the foreground without sacrificing other issues.

The environmental crisis is depicted as a "point of entry" into the politics of Los Angeles, which, according to the Labor/Community Strategy

Center, "demands popular control of corporate decision making for workers and communities" (Mann 1993). There is a continuity between the work on the plant-closing campaign in Van Nuys in the early 1980s and the issues of the environment that are being raised by the center more than a decade later.[23] In its campaign in Wilmington, the center demanded, "communities and workers have a right to decide how polluting industries will behave in their neighborhoods" (Labor/Community Strategy Center 1994a).The framing of the environment has also involved seeking to build a sense of shared interests between workers and community residents around this theme of accountability.

The Labor/Community Strategy Center is uniquely attuned to the concerns of workers, and particularly to the racial and ethnic diversity of the working class in Los Angeles. "In today's society, large corporations are operating with willful disregard for our health, our safety and even our opinions. Thus, the movement for clean air in Los Angeles is really a fight over whether the majority of the people can take control of economic decisions that are presently made by a minority. It will involve new laws, new organizations, new concepts of worker and community rights, and a new social movement demanding a radical redefinition of how this country is owned and operated. That is the challenge, and the fight for clean air is just one critical point of entry" (Mann and the Labor/Community Watchdog Organizing Committee 1991, 72). This framing of the environment focuses to a large part on this issue of production.[24]

By framing its work with the environment in Wilmington and elsewhere in Los Angeles as issues of community accountability, the Labor/Community Strategy Center has been able to bring together the health problems of workers with environmental problems in Los Angeles in a way that has not been successful in many working-class communities. "The environmental crisis and its resolution hinges on production and who controls it" (Mann and the Labor/Community Watchdog Organizing Committee 1991, 47). Not only is the accountability of industry to nearby communities connected with the environment, immigration, education, affirmative action, health care, and economic empowerment are connected with it as well.

In this regard, few issues are as crucial to the center as the racial and class polarization in the city. It may be necessary, the center explained, to " 'isolate' chemical factors from socioeconomic factors" in medical or scientific research (Mann and the Labor/Community Watchdog Organizing Committee 1991, 34). In the "real world," however, explained the group, "they can't be separated." "From an organizer's point of view," concluded the center, "it is precisely the interconnectedness of poor hygiene, poor nutrition, smoking, frequent infections, smog, and air toxins that creates the lethal living conditions for so many low-income, minority, and working-class people. And it is precisely this interconnectedness that creates the potential for a potent movement to change those unbearable conditions" (Mann and the Labor/Community Watchdog Organizing Committee 1991, 34). After a decade of experience with the United Auto Workers and energized by the organizing of the Concerned Citizens of South Central Los Angeles, the Labor/Community Strategy Center has framed an environmentalism that it is taking to the heart of Los Angeles' most impoverished, polarized, and racially diverse neighborhoods.

The Labor/Community Strategy Center has worked to move beyond a narrow understanding of the environment. This work, writes Eric Mann, is not simply "a battle against chemicals but a kind of politics that demands popular control of corporate decision making on behalf of workers and communities" (Mann 1990b, 60).[25] The group also embeds its work with the environment in its organizing to rebuild the economy of Los Angeles. With its recognition that environmental problems are political issues rooted in industry's lack of accountability to workers and community residents alike, the center is framing the environment in some of the nation's poorest communities.

LABOR AND THE ENVIRONMENTAL JUSTICE MOVEMENT

The Labor/Community Strategy Center in Los Angeles and the Oil, Chemical and Atomic Workers in Geismar are paving the way in which the environment in working-class communities can be brought together with workplace and union concerns. Together, they are recasting the environment in terms of workplace health and the accountability of industry to workers and local residents alike. There are certainly parallels in the hazardous waste movement and groups such as the Citizen's Clearinghouse for Hazardous Wastes, which has had a large number of working-class participants.[26]

In his work on the hazardous waste movement, Ken Geiser has written of similar experiences in the early 1980s, as large numbers of working-class communities began to draw attention to the environment. "Because the movement is so tightly rooted in the immediate experience of people's community and family life," writes Geiser, "it has an urgency and a concreteness that is incredibly compelling. For these new 'environmentalists' the environment is not an abstract concept polluted and decreasing in beauty or scenic value. For many it is something which has already exposed them to hazards which are debilitating them and hastening their deaths" (Geiser 1983, 40). Much the same can be said for the involvement of labor unions with the environmental justice movement in Los Angeles, Louisiana and elsewhere.

The history of industrial unionism is also crucial to the framing of the environment by the Oil, Chemical and Atomic Workers and the Labor/Community Strategy Center. In its publications, Local 4-620 points to the history of its national union and its involvement with workplace safety and occupational health. The Labor/Community Strategy Center in Los

Angeles also grounds its work in the history of industrial unionism, particularly that of the United Auto Workers, with its tradition of progressive activism.

For years, working-class communities have played an important part in the efforts throughout the country to reduce hazardous wastes. It is no coincidence that the struggle against hazardous wastes in Love Canal in the late 1970s was a watershed that brought the environmental threats to millions of working people to national attention. The images of working families and children being poisoned by hazardous wastes in the backyards of their homes resonated with the hardships of blue-collar life for millions of people. In much the same way, the Labor/Community Strategy Center and the Oil, Chemical and Atomic Workers Local 4-620 have played upon these familiar images of working-class life in their framing of the environment.

NOTES

1. "While the vast array of aerospace plans and oil refineries provide jobs for a well-paid industrial workforce, large segments of Los Angeles' electronic, chemical, furniture, solvent processing, hotel, restaurant, and garment industries are built on a foundation of low-paid, and often non-union, labor" (Mann and the Labor/Community Watchdog Organizing Committee 1991, 8).

2. "The most dangerous work," explains the Labor/Community Strategy Center, "is reserved for low-paid, minority, and often immigrant, laborers" (Mann and the Labor/Community Watchdog Organizing Committee 1991, 29).

3. The main slogan of the plant-closing campaign was "The future of General Motors Van Nuys, it's not just for management to decide. Workers and communities demand a voice!" (Mann and the Labor/Community Watchdog Organizing Committee 1991,55).

4. "It is the working class that, by its very definition, works with toxic chemicals, and whose members are on the front lines of the battle. It is a rare situation when an industrial plant is poisoning the general population with chemical emissions and the workers in the plant are not being impacted, usually even worse than the general population, by those same chemicals" (Mann and the Labor/Community Watchdog Organizing Committee 1991, 28).

5. "The struggle to clean up L.A.'s air is not a luxury, it is a life and death necessity. If we don't eliminate the toxic chemicals in our air, they will eliminate us" (Mann and the Labor/Community Watchdog Organizing Committee 1991, 6).

6. In its study, *L.A.'s Lethal Air*, the Labor/Community Strategy Center explains, "Los Angeles County reflects the demographics of polarization: great concentrations of corporate and private wealth, a rapidly shrinking lower middle class and unionized working class, and a growing army of the working poor, the unemployed, the homeless, and the criminally employed" (Mann and the Labor/Community Watchdog Organizing Committee 1991, 6–7).

7. The group began its work with the environment because it was "critical of what we felt were the class and race insensitivities of some of the more established environmental organizations" (Mann and the Labor/Community Watchdog Organizing Committee 1991, 61).

8. As the Labor/Community Strategy Center concluded in its open letter to Wilmington residents, "we know that we face dangers from other refineries and chemical plants [in Wilmington] and see the struggle for corporate responsibility from Texaco as the first step toward an overall plan for the protection of community health and safety. Success in this struggle will only be achieved through community involvement. Come join us in the fight for your family's health!"

9. As the Labor/Community Strategy Center explained, "medical treatment must be offered at a central location at hours convenient for working persons and Texaco must continue to accept claims to cover the costs for medical treatment beyond what is currently an unreasonable deadline" (Labor/Community Strategy Center 1994a).

10. Wrote Labor/Community Strategy Center activist Kikanza Ramsey, "we demand medical rescreening to investigate any long-term health effects and address why [local residents] still are feeling effects such as unstable nerves, hearing trouble, and loss of taste and smell" (Ramsey 1993).

11. In its *An Open Letter to the Workers at Texaco's Wilmington Refinery from the Labor/Community Watchdog* (1994a), the Labor/Community Strategy Center explained "Texaco must stop terrorizing the community by asking for green cards before medical services are provided, as some community members have experienced. This racial insensitivity is not isolated, but rather is a continuation of treatment toward Spanish speaking residents that started with the lack of an 800 [medical care] claims [telephone] number in Spanish."

12. In another flyer, the group emphasizes "we are fighting to clean up L.A.'s air, protect the health and safety of workers and communities from chemical accidents, and reduce the millions of pounds of health threatening pollution produced by the chemical and automobile industries" (Labor/Community Strategy Center 1993b).

13. "During the 1930s, it was not Franklin D. Roosevelt as much as union organizers from the Congress of Industrial Organizations (CIO) along with other progressive and left activists, who provided leadership and organized to the unemployed, to dispossessed farmers, to Southern black and white sharecroppers, and to World War I veterans who had been denied their benefits. This broad social movement from the bottom up was the driving force behind what later came to be called the New Deal" (Mann and the Labor/Community Watchdog Organizing Committee 1991, 72).

14. "Our organizing work is driven by a vision of direct democracy. We are attempting to build a powerful and democratic organization, beginning with dozens and growing to hundreds and thousands, with the explicit goal of challenging the domination of our lives by the DuPonts, Dows, Chevrons, Unocals, and General Motors of this world" (Mann and the Labor/Community Watchdog Organizing Committee 1991, 72).

15. Writes Eric Mann, "narrow demands for reduced toxic chemical use that do not take into account job costs inevitably end up scapegoating industrial workers and driving them even further into the arms of management" (Mann 1992, 16). As the Labor/Community Strategy Center explained elsewhere, "from our perspective, we are most interested in raising this demand [for worker education and retraining funds] in conjunction with a campaign to temporarily shut down a toxic corporate polluter, during which time workers would be paid while the company was compelled to bring in non-polluting technology" (Mann and the Labor/Community Watchdog Organizing Committee 1991, 66).

16. "If we have any hope of constructing a society that is based on industrial democracy and environmental safety, we need a strategy that targets corporate pro-

duction" (Mann and the Labor/Community Watchdog Organizing Committee 1991, 46).

17. As a leading figure with the Labor/Community Strategy Center explained, "the 'environment' is not just a white people's issue or a middle class issue but is our issue as well" (Mann and the Labor/Community Watchdog Organizing Committee 1991, 62).

18. Parking for municipal employees and others had been a negotiated benefit for more than a decade prior to this controversy. Researchers with the Service Employees International Union Local 660 in Los Angeles "uncovered documents proving that the county's real intention was to drive county workers from those lots in order to deliver the lots as part of land packages to large developers, in return for ground rents that would go into the county treasury. The plans of these developers to construct hotels and high rises on the sites would, in fact, increase the number of cars in the [downtown] area by four-hundred percent and thereby dramatically increase the air pollution as well" (Mann and the Labor/Community Watchdog Organizing Committee 1991, 51).

19. As part of its efforts to frame an environmentalism that is part of life for working-class communities, the Labor/Community Strategy Center has devoted effort to the study of environmentally sustainable technologies that would rebuild the economy of Los Angeles.

20. "For low income communities and communities of color, replacing the congestion and pollution of auto traffic with a public transportation system that is organized around indigenous needs can become an exciting element of an overall plan for community revitalization" (Mann and the Labor/Community Watchdog Organizing Committee 1991, 69). The Labor/Community Strategy Center is "studying the most environmentally benign bus and engine designs and proposes that buses be built in Los Angeles' own rust belt by workers from the closed-down Ford Pico Rivera, Bethlehem Steel, Goodyear, Firestone, and General Motors Southgate plants, and the threatened GM Van Nuys plant" (Mann and the Labor/Community Watchdog Organizing Committee 1991, 69).

21. "Since we believe that workplace and community organizing, direct action and popular control of corporate production must set the terms of the debate, there is a need to change the social composition not only of the troops but of the leaders" (Mann and the Labor/Community Watchdog Organizing Committee 1991, 58).

22. The Labor/Community Strategy Center is "placing environmental activism in the context of rebuilding the progressive movement" (Mann 1993). The group sees itself as making "an effort to revitalize the progressive movement in Los Angeles, integrating issues of the environment, racial equality, union organizing, women's rights, community empowerment, world peace, and international solidarity into an overall movement for economic and political democracy. Toward this goal, we study the history of social movements and draw on their lessons in an effort to creatively apply them to our work" (Mann and the Labor/Community Watchdog Organizing Committee 1991, 72).

23. The plant-closing campaign "was over the political and economic process by which decisions over plant closings were made: who makes the decisions and who suffers the consequences?" (Mann 1987, 109).

24. The environment is linked to "the social control of production" (Mann 1990a).

25. Writes Mann, "the cutting edge of environmental organizing depends on direct, multifaceted campaigns against corporate toxic polluters, the Unocals, the General Motors, the Dows, and the Chevrons, involving tactics such as demonstrations, public education campaigns and large-scale consumer boycotts demanding

the immediate cessation of a particular product or process and its replacement with an environmentally sound one" (Mann 1990a, 270).

26. The Love Canal Homeowners Association and the work of Lois Marie Gibbs arose from a working-class culture of political activism and organizing that framed its concerns with hazardous wastes in terms of a homeowners' association concerned with property values as well as matters of health.

6

"More Political Than Chemical": Toward a Cultural and Political Perspective of the Environmental Justice Movement

We are caught in an inescapable web of mutuality, tied to a single garment of destiny. Whatever affects *one* directly affects *all* indirectly.

—The Reverend Martin Luther King, Jr.

The involvement of African-Americans, Latinos, Asian Pacific, and Native Americans, working people, and the poor in the environmental justice movement reflects a weaving together of the environment with the inequalities faced by these groups. The environment is incorporated into organizing by groups with years of experience in these communities, who have recast their definition of the environment to connect it with housing, municipal services, workplace safety, health care, and property taxes.

The Gulf Coast Tenants Organization, the Labor/Community Strategy Center, the Oil, Chemical and Atomic Workers Local 4-620, and the South-West Organizing Project each incorporate the environment into their work. The framing of the environment by each of these groups took place in the context of organizations that had already been working with these communities for years.

Economically, many African-Americans, Latinos, Native Americans, and other minority groups still lag behind much of the rest of society. The widening gap in wealth and income at the end of the century is only deepening disparities in health and the environment. In many communities, the decline of traditional manufacturing employment in the face of the globalization of markets abroad and technological change at home is rendering environmental problems trivial by comparison. Yet here, on the front lines of some of the most difficult economic realities at the end of the century, communities with a history of estrangement from the postwar environmental movement are reshaping their understanding of the environment and building a remarkable new movement for social change.

RESOURCE MOBILIZATION, POLITICAL PROCESS THEORY AND A CULTURAL THEORY OF SOCIAL MOVEMENTS

The emergence of political movements is a perennial area of interest in the study of history, politics, and sociology. Much of this research focuses on the difficulties faced by groups in their mobilization efforts. Although there is a widely cited literature in the study of movements, there are clearly other factors which should be taken into consideration, especially the intersection of language and culture as it relates to political activism.

While this literature is useful for understanding the circumstances that contribute to the emergence of a movement, it is also necessary to understand the way in which the leaders in a movement use language to frame problems for participants. How are problems depicted by the leaders of a movement? How do those involved in a movement come to see their own circumstances as connected with the circumstances of others? What is the new identity that individuals create for themselves as participants in a movement? And how does language itself shape the possibilities for social change?

The theoretical perspective of framing is offered in this book as necessary for understanding this politicization in a movement. Framing is defined in this book as a way of simplifying a problem, of rendering it in the most understandable terms or language for a particular group. The theoretical perspective of framing is necessary for understanding the way in which the leaders of a movement carefully word the problems being dealt with in such a way as to mobilize participants.[1] The perspective of framing holds that the leaders in a movement do not merely address a problem on its own terms but fit it in to their own beliefs and their own language.

For many years, the ability of language to shape political activism has been an underappreciated aspect of the study of movements. Yet, as argued in this book, few movements can be understood without an appreciation of what is referred to here as collective identity.[2] The collective identity of a movement is crucial to its participants.[3] Involvement in a movement engenders new forms of self-understanding for activists, often reflective of their shared experiences of activism.[4]

A political movement is a shared experience. Individuals form an identification with others in the context of their involvement with a movement, building new bonds with each other through the struggle and the sacrifice that comes with such an involvement. In a very real sense, a collective identity can be thought of as the new self-understanding people shape from their own involvement in a movement.

The collective identity of a movement is powerful to the extent that it merges with the existing identities of individuals and at the same time expands these to create new, even more desirable identities.[5] Debra Friedman and Doug McAdam maintain that involvement in a movement "makes it more likely that the individual will value the identity of 'activist' and choose to act in accordance with it" (Friedman and McAdam 1992, 170).[6] As an example of this, Friedman and McAdam point to the early civil rights movement, which built upon the religious commitments among African-Americans. The civil rights movement "came, in many Southern towns, to control access to a role that was highly prized. Blacks in those towns, to retain their status as Christians, added civil rights activities to their other Christian duties" (Friedman and McAdam 1992, 163). This emergent collective identity in the civil rights movement and its idea of nonviolent civil disobedience was intermeshed with African-American churches.[7] This same process has played itself out in nearly every movement, insofar as existing identities have been refashioned around new roles or self-understandings.[8]

The environmental justice movement, it is argued here, has formed a distinctive collective identity for its participants through their activism and organizing around environmental issues. It is a new collective identity that encompasses nearly every prior kind of organizing by subsuming it under a struggle to protect an environment defined as "where we live, work and play." Even as activists continue to see themselves as neighborhood organizers or as trade unionists, they also come to see themselves as environmental justice activists.

"THIS IS JUST ONE MORE THING": FRAMING THE ENVIRONMENT AS AN INJUSTICE

Some of the most fascinating literature in the study of political movements covers the extent to which new issues or concerns are grafted onto such movements. "Successful movements usually do not create attractive collective identities from scratch," write Debra Friedman and Doug McAdam. "Rather, they redefine existing roles within established organizations as the basis of an emerging activist identity" (Friedman and McAdam 1992, 162).[9] As argued in this book, framing is not only a way of explaining the problems facing a group or community, but of grafting new issues onto the concerns of these groups.

The Gulf Coast Tenants Organization and SouthWest Organizing Project each depict the environment in language that renders it part of the inequalities their communities have had to endure for decades and even centuries. The language of racial injustice is commonplace in the framing of the SouthWest Organizing Project and Gulf Coast Tenants Organization. A language of injustice also intermingles with a wider appeal to the religious faith of many groups in the environmental justice movement.[10] Each of these groups depicted their communities as having suffered pollution from outside businesses, industry, government or other forces. Many identified their communities as dumping grounds or what the SouthWest Organizing Project called sacrifice zones. Because of environmental conditions, the Gulf Coast Tenants Organization likened the towns along the Mississippi River to a Third World country. Local 4-620 of the Oil, Chemical and Atomic Workers described the state of Louisiana as a resource colony.

For the Gulf Coast Tenants Organization, the SouthWest Organizing Project and other groups, environmental hazards were seen as products of decisions that placed these communities in harm's way. In its March 1990 letter, the SouthWest Organizing Project reflected that "industrial and municipal dumps are intentionally placed in communities of color" (SouthWest Organizing Project 1990a).[11] In framing environmental problems as related to the systematic siting of facilities such as incinerators, the leaders of the SouthWest Organizing Project and the Gulf Coast Tenants Organization connected with a history of discrimination.

Histories of discrimination and inequality are used to provide a way of thinking about the environment. For the Gulf Coast Tenants Organization, the history of the plantations was resonant for the public housing projects and towns along the Mississippi River.[12] The Oil, Chemical and Atomic Workers Local 4-620 similarly drew from the history of the plantations. Local 4-620 had a sizable proportion of African-American members and its leaders wanted to build a connection with the residents in the towns surrounding the BASF plant, even though the union was more ready to connect the environment with property tax and municipal infrastructure concerns.

In even the most influential publications in the environmental justice movement, there is a recognition that while environmental racism has been the most common way of framing the environment, the quality of affordable housing, property values, and impoverishment are also important to these groups in terms of explaining the incidence of environmental problems.[13] The lack of zoning regulations in less privileged communities may influence the decisions of industry to locate in these communities.[14] Industries may also locate in the vicinity of transportation routes, such as interstate highways and railroads, which already tend to have lower property values.[15] In short, the worst hazards from industries tend to be found in these communities for a number of reasons.

"WE DON'T TALK ABOUT ENVIRONMENTALISM": FRAMING THE ENVIRONMENT AS WHERE WE LIVE, WORK AND PLAY

The Gulf Coast Tenants Organization, the SouthWest Organizing Project, the Labor/Community Strategy Center, and the Oil, Chemical and Atomic Workers Local 4-620 each went to considerable lengths to frame the environment in terms that were consistent with their previous work.[16]

The SouthWest Organizing Project and Gulf Coast Tenants Organization framed the environment as "where we live, work and play," while the Oil, Chemical and Atomic Workers Local 4-620 and the Labor/Community Strategy Center used a language of workplace rights and accountability to workers and local community residents.

The widely cited understanding of the environment as "where we live, work and play" is a crucial part of framing the environment in Asian Pacific, Native American, African-American, and Latino communities. As shown in the preceding chapters, each of these groups sought to change the language or wording used to describe the environment so as to make it more a part of their previous work. The wording of the environment was cast in the widest terms possible.

The Labor/Community Strategy Center and SouthWest Organizing Project focused on the way in which the "narrowness" of previous understandings of the environment have left out a host of concerns. Their new definition of the environment was premised on the incorporation of many different issues, in contrast to what they characterized as a "narrow" or "single-issue" environmentalism from the postwar environmental movement.

The commonplace reference to the environment in quotation marks or italics in the literature of the SouthWest Organizing Project, the Labor/Community Strategy Center, and the Gulf Coast Tenants Organization reflected the extent to which the environment has been a problematic concept. For the Gulf Coast Tenants Organization, the problems that prompted the involvement of local residents included the accidental releases of hazardous materials along the Mississippi River. This coupling of the conditions in public housing with accidents in the chemical industry was a way of framing the environment to have a connection with the organizing of the tenants' association, for whom the health concerns of public housing residents had always been a priority.

For Oil, Chemical and Atomic Workers Local 4-620, property taxes were seen as some of the most immediate concerns of the towns along the Mississippi River. With its formation of the Louisiana Coalition for Tax Justice, Local 4-620 was able to expand its organizing to include larger questions regarding a statewide policy in Louisiana of tax incentives for the chemical industry.

The environment was part and parcel of the many other problems in Los Angeles, Albuquerque, Geismar, and New Orleans. That these groups repeatedly used the term environment in quotation marks reflected the work of these groups to create a new definition of the environment as distinct from that of the postwar environmental movement. Far from being a matter of semantics, this framing of the environment had implications for day-to-day organizing. The flyers, leaflets, books, and posters of these groups were replete with references to this "new definition" of the environment. The leaders of these groups acknowledged that they had carefully chosen the wording of their new definition of the environment in order to cast it in terms that appealed to the most immediate concerns of local residents. In their work in Los Angeles, organizers with the Labor/Community Strategy Center acknowledged that they had all but avoided use of the term "environment" in their literature.

The groups in the environmental justice movement were also reluctant to describe their work as part of the postwar environmental movement. There was a hesitance on the part of these groups to change their names or otherwise revise their mission statements, even though the environment had come to comprise a larger portion of their work. The labor unions working with the environmental justice movement tended to depict the environment in terms of "rights for workers," "occupational safety," or the "accountability" of industry to workers and community residents. Only after assuring communities and workers alike of their desire to keep local businesses open did these groups begin to address concerns in terms of the environment.

The Gulf Coast Tenants Organization, the SouthWest Organizing Project, and other groups each sought to expand their communities' understanding of the environment. The environment was understood to include housing, education, and property taxes. It was recast as an integral part of many other problems in these communities, whether the lack of a municipal infrastructure in the towns along the Mississippi River or the lack of accountability of industry to residents in Los Angeles.

PROTEST TRADITIONS AND THE ENVIRONMENTAL JUSTICE MOVEMENT

The theoretical perspective of framing in political movements must address the way in which movements are shaped by the history of previous movements. As argued here, the environmental justice movement has been understood through the historical experiences of racism and inequality. In poor, working-class and African-American and Latino communities, these environmental justice groups were more likely to trace their history to the civil rights, farmworkers, or labor movements. Thus, the "freedom movement," as it was often described by public housing residents, played a significant part in the work of the Gulf Coast Tenants Organization. For the SouthWest Organizing Project, this history manifested itself in a connection with the United Farmworkers.

Much like the work of these community groups, the unions and working-class communities were also found to have connected their involvement around environmental issues with a wider history of labor organizing. Oil, Chemical and Atomic Workers Local 4-620 drew from the history of their union's involvement with environmental health as a part of a wider tradition of labor activism. Additionally, the Labor/Community Strategy Center also used the history of unionism and labor organizing in the 1930s for its framing of the environment.

The Gulf Coast Tenants Organization, the SouthWest Organizing Project, and the Labor/Community Strategy Center were each found to be reluctant to consider their work with the environment as part of the postwar environmental movement. Throughout the publications of the Gulf Coast Tenants Organization and the SouthWest Organizing Project, the environmental movement was framed as "elitist," "European," "luxury oriented," "narrowly focused," and "legalistic."

The SouthWest Organizing Project and Gulf Coast Tenants Organization described their work as a "new social justice" or "new civil rights" movement, as a part of what is referred to as "the struggle for a more just and democratic South in the 1990s," according to the Gulf Coast Tenants Organization. Each group underscored the importance of their continuity with the civil rights movement. The Labor/Community Strategy Center saw itself as part of a struggle for what it called "industrial democracy," which had roots in the labor movement of the 1930s. The United Auto Workers and the United Farm Workers, more than the Sierra Club or the Wilderness Society, framed the environment in terms of these histories of struggle.

For the SouthWest Organizing Project, there is a connection to the history of the Southwest, and to the land itself. The history of resistance against the encroachment on land and its impact on indigenous cultures in the Southwest is uppermost for the group. As "a contemporary struggle with very old roots," the project has reconnected the history of struggle against encroachment on the land with the work to solve environmental problems. The prominence of the civil rights, labor union, and tenants' movements in the framing of the environment by the SouthWest Organizing Project and Gulf Coast Tenants Organization, then, reflect the extent to which these movements remain enduring parts of a history for these groups.

RACISM, POVERTY AND THE STRUGGLE FOR A NEW ENVIRONMENTALISM

The work of the Gulf Coast Tenants Organization, the SouthWest Organizing Project, the Oil, Chemical and Atomic Workers Local 4-620, the Labor/Community Strategy Center, and other groups in the environmental justice movement is extremely important, because it comes at a time when the nation's largest environmental groups have experienced a decade of challenges from communities throughout the country. From the controversies over logging in the forests of the Northwest to the wise use movement of the Western states, the last decade has seen challenges to the postwar environmental movement. In settings as varied as a public housing project in Louisiana and a windswept field in Montana, local residents have challenged what they depict as the "elitism," "arrogance," and "indifference" of environmental groups in their communities.

The environmental justice movement and its framing of the environment around the inequalities of racism is only one example of the way in which understandings of the environment are connected with the experiences of different groups. It is important, then, to recognize the many different ways in which these groups are involved with environmental problems in the surroundings most familiar to themselves, their families, and their friends. As with most problems, the environment is perceived differently by groups according to their immediate circumstances.[17]

In its framing of environmentalism as well as its ubiquitous quotation marks in written discussions of the environment, the work of such groups as the Labor/Community Strategy Center and the SouthWest Organizing Project has revealed the environment as a problematic concept for workers, the poor, and people of color. The environment has a diffuse character, so that the experiences of different groups can be brought to the framing of the environment. Every effort to frame the environment has to take into account the language that groups bring to their involvement with the environment.[18]

For example, the most important aspect of the work of the SouthWest Organizing Project, the Gulf Coast Tenants Organization, the Labor/Community Strategy Center, and other groups is their connection of the environment with the inequities of racism and impoverishment. Thus, the Gulf Coast Tenants Organization, the SouthWest Organizing Project, the Oil, Chemical and Atomic Workers Local 4-620, and the Labor/Community Strategy Center are forging what I call a competing narrative of environmentalism, one that brings the experiences of inequality and impoverishment for working-class and people of color to the foreground.[19] This narrative offers its own terms and even its own history of the struggle to protect the environment, drawing from the civil rights, tenants, farmworkers, and labor movements.

More than twenty years ago, Daniel Zwerdling reflected on the postwar environmental movement and its failure to connect with the poor, in particular to those in the nation's inner-city neighborhoods. Even though concerns such as the Vietnam War and the deterioration of the nation's inner cities had been found in the literature of the Earth Day events in April 1970, these concerns had fallen by the wayside amid the work of newly founded groups such as the Environmental Defense Fund, the Environmental Action Foundation, and the Natural Resources Defense Council. The mainstream environmental movement was seen as disconnected from the problems of the nation's poor, explained Zwerdling. "If the environmental movement is going to survive, it must revamp the most fundamental notions of what it is all about. For individual workers and the poor, the environment is not three hundred miles away and six hours by car, it begins every morning at home and when the first whistle blows on the job"

(Zwerdling 1973, 25). The environmental movement lacked an immediacy that could connect with problems of racism and poverty.

Twenty years later, the Gulf Coast Tenants Organization, the SouthWest Organizing Project, the Labor/Community Strategy Center, the Oil, Chemical and Atomic Workers Local 4-620, and many other groups have made an effort to frame definitions of the environment that incorporate the conditions of racial inequality, impoverishment, and the environment. Whether in African-American churches along the Mississippi River or labor union meeting halls in Los Angeles, groups have fashioned a new language of environmentalism, one that is broad enough to encompass much of their previous political organizing.

In their work, the Gulf Coast Tenants Organization, the SouthWest Organizing Project, the Oil, Chemical and Atomic Workers Local 4-620, and the Labor/Community Strategy Center were found to have connected the environment with many of the other problems facing their communities. What is learned in this book is that people are not passive recipients of terms like "the environment." They actively reshape them to fit the circumstances of their lives. They rework these concepts and refract them through their own experiences. Language is not frozen in a movement but is open to struggle, contention, and conflict.

Among the most significant accomplishments of the environmental justice movement has been its broadening of the language used to refer to the environment to include issues of racism, poverty, and the legacy of injustice. The importance of language to the environmental justice movement should in no way discount the significance of the socioeconomic and political dimensions of these environmental struggles. This book, however, would expand the research on the incidence of environmental problems by recognizing the significance of language and the new vocabulary of race, rights, and justice that has had an impact on this unique brand of environmentalism. The environmental justice movement is the most significant recent contribution to this nation's struggle to protect the environment, and promises to have a lasting effect on the ways in which policy makers, businesses and industry, and citizens alike think of the environment in the years to come.

NOTES

1. The research on framing goes back to Erving Goffman's work on frame analysis, insofar as frames are understood as the conceptual filters used to identify, understand, or otherwise make sense of problems (Goffman 1974).

2. In recent years, research has focused on "the central importance [movements] attached to the creation of new collective identities as a fundamental goal of the movement" (McAdam 1994, 50). As Doug McAdam concludes, "social movements have always served this function, whether it was an explicit goal of the movement or an unintended consequence of struggle" (McAdam 1994, 50).

3. As David Snow and Pamela Oliver have written, "given the centrality of the concept of identity to much theorizing and research in social psychology and the fact that participation in various crowd and movement activities can call into question and modify existing identities as well as provide new ones, it is hardly surprising that the identity concept would find its way into the study of social movements" (Snow and Oliver 1993, 23).

4. Theoretical research on movements, according to Robert Benford, should study "how collective identities are constructed in the course of participation" (Benford 1993a, 210).

5. To embrace the collective identity of a movement is "to claim for oneself a desired social attachment and new sense of identity" (Friedman and McAdam 1992, 157).

6. As Pamela Oliver has written, "movement activists come to value the image of themselves as activists as an end in itself" (Oliver 1989, 17).

7. As Aldon Morris and his colleagues have written, "the preexisting attitudes associated with black religion and the new attitudes and knowledge associated with nonviolent resistance combined to produce the consciousness that gripped black Americans who participated in the civil rights movement" (Morris, Hatchett and Brown 1989, 285–286).

8. In his work on the peace movement, Robert D. Benford has discussed this in terms of the new self-identities fashioned by activists involved in civil disobedience. "Having been arrested for 'Doing CD' tended to confer status honors upon the arrestee, honors that were often collectively interpreted as more than outweighing the temporary inconvenience of detention. To neglect the processes by which such experiences become imbued with meaning leads to reifications and distortions of the actors' realities" (Benford 1993a, 210).

9. As Debra Friedman and Doug McAdam maintain, "people affiliate with groups for a variety of reasons, but they are not about to do so if the group's identity is incompatible with their image of themselves" (Friedman and McAdam 1992, 165)

10. "Israel's young warrior, David, wrote with God's inspiration the twenty-fourth Psalm: the earth is the Lord's and the fullness thereof, the world and they that dwell therein. This doesn't mean that the earth belongs to Cargill and Marathon. The magnificence of God's creation is being torn apart by profit makers so fast that we are globally asking the question: how much longer do we have to stop these companies that ruin the world before it is too late?" (Gulf Coast Tenants Organization 1993b, 1).

11. Explained an activist involved in the SouthWest Organizing Project, "Communities inhabited by people of color are targeted for industries which would not be allowed in other neighborhoods" (Martinez 1991a, 4).

12. For many, the most devastating consequence of the environmental problems along the Mississippi River is not merely the loss of communities themselves but an entire way of life and a culture that is being destroyed. As the Gulf Coast Tenants Organization explains of the petrochemical industry and its impact on communities along the Mississippi River, "more than a century old slave communities are being uprooted and a culture that has helped keep African-Americans in this region healthy is threatened" (Gulf Coast Tenants Organization 1992).

13. "Many interrelated factors contribute to today's situation," writes Luke Cole of the California Rural Legal Assistance Foundation, "including industry's tendency to seek inexpensive land in low-income neighborhoods as well as poor people's lack of political and economic power in resisting such intrusions" (Cole 1992, 628).

14. As Bunyan Bryant of the University of Michigan's School of Natural Resources and Environment has reflected, "in a black neighborhood, the fine is a slap on the wrist, the cost of doing business. That in effect encourages industries to pollute in minority neighborhoods" (Bryant 1993, 10).

15. There is preliminary evidence from some researchers that hazardous waste sites, particularly illegal and abandoned waste sites, are often located along highways and railroad corridors in metropolitan areas that may already have lower income residents.

16. Giovanna Di Chiro has written, "it would be more accurate to regard environmental [justice] activists as the 'new' civil rights or 'new' social justice activists, since many of the prominent organizers affirm their roots in and political continuities with the social justice movements of the sixties, including the civil rights, welfare rights, and labor and farm worker movements" (Di Chiro 1992, 97).

17. "As [is true of] all other social problems, the environmental crisis is perceived differently by different social groups according to the group's socioeconomic position, its ideology and culture, and the other social problems the group faces" (Environmental Careers Organization 1992, 69).

18. As David Harvey has written, "each and everyone of us is situated in an 'environment' and all of us therefore have some sense of what an 'environmental issue' is all about" (Harvey 1993, 2).

19. Many other groups involved with or a part of the environmental justice movement went to the same lengths to connect the environment with other often disparate issues. In 1989, writes Randy Stoecker, the Citizen's Clearinghouse for Hazardous Wastes, a group with a long-time connection to the environmental justice movement, wrote and distributed a written endorsement of a national housing march entitled *Ten Points of Common Interest Between the Environmental Justice Movement and the Homeless Movement.* The ten points emphasized a connection between the struggles to protect the environment and the struggle to end homelessness, explaining "some people, especially those whose homes are contaminated by pesticides, have abandoned those homes and become homeless" (Stoecker 1992, 247).

Conclusion: "The Environmental Crisis Impacts Us All but Not Equally": Language, Ideology and the Contested Narrative of the Environment in the Environmental Justice Movement

An all white environmental movement is in itself an environmental crisis.

—The Reverend Jesse Jackson

The scholarly interest in the environment has expanded at a breathtaking pace during the past decade. History, sociology, anthropology, philosophy, and even art and literature are a part of a blurring of disciplinary genres in which an understanding of the causes and the consequences of environmental problems at the end of the twentieth century is being sought. The environment is a presence in movies, television, and film as well as magazines, books, and even children's cartoons.

The environment is a commonplace theme in television, movies, newspapers, and films. It is a cultural phenomenon, perhaps one of the most important of the postwar period, and certainly one that shapes nearly every aspect of our lives today. Environmental-friendly products make their way into the supermarkets and homes of millions of Americans. The environment is so much a presence that it even shapes the language we use to reflect on ourselves and our social lives.

The language used to refer to the environment is rich with cultural connotations, laden with the history of social relations and struggles between peoples. The way we think, speak, and write about nature is as important for telling us about ourselves as it is about the environment itself. Nature, writes Raymond Williams, "contains, though often unnoticed, an extraordinary amount of human history" (Williams 1984, 82). It shapes how we view ourselves and the world around us and reflects back at us our own beliefs, biases, and ways of thinking. In one of the most widely influential formulations, Raymond Williams writes that the ideas of nature "are the projected ideas of men" (Williams 1984, 82). The ideas of women and other groups, it might be emphasized, are crucial to ideas of nature as well, although one might hasten to add that Williams' formulation is itself revealing of the engendering of nature and the extent to which the environment always and everywhere reflects the partiality of the artists and authors who write about, paint, and photograph it.

It is clear from the previous chapters in this book that the groups in the environmental justice movement draw from a distinct set of experiences as a way of relating to and understanding the environment. The Labor/Community Strategy Center and the Oil, Chemical and Atomic Workers Local 4-620, organizing sometimes literally in the shadows of industry, grappled with the anguish of weighing their own employment against the threats to the environment. The Gulf Coast Tenants Organization struggled for affirmative action in the hiring policies of the chemical plants along the Mississippi River even as it battled against the pollution from these same plants.

The environment is always situated in people's places of residence. An African-American woman in the public housing projects of the south side of Chicago is going to think much differently (though perhaps no *less*) about the environment than a lawyer with the Environmental Defense Fund in New York. A Laotian fishing family in Oakland is much differently situated to think about the environment than an urban planner in San Francisco or a graduate student shopping at a grocery cooperative in Berkeley. The environment, in short, is a cultural formation, an expression of social relations. History, too, bears the imprint of nature.

THE ENVIRONMENTAL JUSTICE MOVEMENT AND A HISTORY OF ENVIRONMENTALISMS

Environmental history is concerned with the intersections of history and nature. From the writings of Aldo Leopold on the sand counties of Wisconsin to the scholarship of William Cronon on the landscape of New England, scholars have thought about the history of the land and the places around them. Historians now routinely take into consideration the ways in which the environment is a part of social, economic, and even political change. Even with the widening interest in environmental history, however, much of this scholarship is limited to a history of more affluent groups rather than to one of inequalities that historically have made some communities more vulnerable to environmental hazards than others.

Through a growing history of more diverse environmentalisms, scholars are becoming more attentive to the fact that in every historical period, there are different, conflicting, and sometimes contradictory understandings of the environment. This is especially crucial in light of the emerging involvement of African-Americans, Latinos and other groups with the environmental justice movement. This movement shows that it is necessary to reconstruct narratives about the environment not merely from the perspective of one particular group, but from many different groups.

Environmental historians must continue to widen their historical perspectives to include not only the European or Anglo-American experiences of the environment, but the experiences of other cultures. In the southwestern United States, countless struggles between preservationists and local residents marked the turbulent history of environmentalism at the turn of the century. A history of the environment in the Southwest is thus both a history of the land and its preservation from despoliation *and* a history of the disownment of the peoples living on these lands. These histories exist alongside one another in the same moment, in conflict with one another in ways that enrich our understanding of the history of the Southwest.

The struggle to write a history of environmentalism is the struggle to recognize the diversity of ways in which different groups have lived with their environment. In his work on Monterey Bay, James O'Connor sheds new light on the actors who are a part of the environmental histories of the coastal region of California, the weekend naturalists and amateur biologists who in their own ways study the wild life and the wilderness around them. From beachcombers to local storytellers, the shores and beaches of Monterey Bay are alive with histories (O'Connor 1995, 21–47). In a different context, the work of historian Andrew Hurley underscores the varied relations with the environment of different racial groups in the industrial setting of the Great Lakes region (Hurley 1995). Hurley's work on the white working-class and African-Americans in the city of Gary, Indiana is yet another example of the way in which historians can incorporate the experiences of any number of different groups to tell their stories about the environment at particular moments in time.

IDEOLOGY AND NATURE IN THE LANGUAGE OF ENVIRONMENTALISM

Nature can be politically reassuring for anyone who wants it to be.

—Andrew Ross

Nature is inseparable from ideology. Indeed, few understandings of ideology exist without reference to nature. "It is one of the tenants of ideology," writes Terry Eagleton, "to naturalize social reality, to make it seem as innocent and unchangeable as nature itself" (Eagleton 1983, 135). "Ideas that draw upon the authority of nature," agrees cultural theorist Andrew Ross, "always have their origins in ideas about society" (Ross 1994). "Facts of nature spawned by the needs of ideology," writes historian Barbara Fields in her work on slavery in the United States, "sometimes acquire greater power over people's minds than facts of nature spawned by nature itself" (Fields 1990, 106).[1] Social relations seem inextricable from nature because the cultural formation of ideology so often reads these relations into nature.

Nature and ideology are inextricably bound; as such, the idea of exactly what and where nature is located is fraught with ideology. Nature, within much of contemporary environmental discourse, is miles from the places where many of the people referred to in this book actually live and work. Nature is found in rowing across a New England lake with a naturalist. It is not found in rowing across a lagoon on the south side of Chicago where the contamination is so great from decades of pollution that the seals of rowboats corrode almost instantly, causing them to sink before technicians can measure the lagoon's level of toxicity. Nature is found with scientists seeking to determine the extent of global atmospheric deterioration. It is not found in the parking lot of municipal employees in Los Angeles fighting against an increase in parking fees for city employees. Nature, everyone knows, is on the lake in New England or the mountainside in Hawaii. Of course the lake and the mountain are nature. No doubt whatsoever. But a municipal parking lot in Los Angeles? Or a paper mill in Savannah? Or a union hall in Central Illinois? Or a wastewater treatment plant in Albuquerque? This surely is not nature, or at least it is far away from what is widely accepted in most scholarship as the environment.

Across the country, universities and colleges, fueled by foundations and grants, and led by faculty, administrators, and students, are now adding environmental studies to their curricula. Without dealing with the inequities in society, though, environmental education is likely to have little impact beyond the classroom or the library. There is a comfortable assurance in the idea that modifying the curriculum in the absence of larger socioeconomic changes will contribute to a resolution of environmental problems. It is a familiar solution for a society unwilling to tackle the social inequalities that so clearly manifest themselves in the environments of our nation's poorest communities.

While suburban elementary school children watch Discovery Channel specials on the tropical rainforests, African-American and poor white families across town live with the realities of hazardous waste and employment in industries that expose them to the worst of chemical hazards. It is in this sense that much of what falls under the rubric of environmental education is in fact inseparable from the ideology of liberal culture, specifically a tendency to collapse the eradication of social problems into education. Merely expanding the curriculum is a substitute for the unpleasantness of political struggle, as if a new reading list in the curriculum is interchangeable with social change. This rests on the erroneous assumption that if we know more we will do more. It is an ideological assumption in liberal culture (and hence an unquestioned and seemingly self-evident proposition, "of course it is better that students learn about the environment. Why would anyone think otherwise?") that tends to obscure the class inequalities that are responsible for what Jonathan Kozol calls the savage inequalities between the wealthy and the poor in terms of both education and the environment.

A glance through the popular titles on the environment speaks volumes to the tendency of these publications to be fit into the culture of consumerism, marketing, and even commuting: *Two Minutes a Day for a Greener Planet; The Green Commuter; Fifty Easy Things You Can Do to Save the Planet; Shopping for a Better World; Fifty Simple Things You Can Do to Save the Earth; The Green Consumer Supermarket Guide; Fifty Simple Things Your Business Can Do to Save the Earth; 101 Ways We Can Make a Difference and Help Save Our Planet*. The titles of these books speak volumes to the tendency of writers to adhere to the ideological parameters of a liberal (even self-help oriented) culture that does little if anything to question the ethics of capitalism itself. With what some might say is misplaced optimism, the EarthWorks Group in Berkeley, California in its *Fifty Simple Things You Can Do to Save the Earth* exhorts individuals to take on what it calls "unbelievably" easy things, reassuring its readers—in a language seemingly out of keeping with the radical spirit nurtured in Berkeley over the years—that there is "no point in letting the news reports and magazine coverage drive you to despair." Never mind the news of oil tanker spills, the illegal dumping of hazardous wastes, or what are characterized as intractable environmental problems, writes Chris Calwell of the EarthWorks Group, "go to it" and protect the environment in "unbelievably easy ways," with an enthusiasm worthy of advertising campaigns for sportswear and athletic shoes rather than as an engaged citizenry tackling a legacy of environmental problems.

Most of the popular books on the environment remain too partial to the consumer ethos. Eradicating any vestiges of class struggle or racial inequality, these works cleanse the environment of its sometimes bitter and hard-fought struggles. They stress the lone shopper asking for paper bags at the supermarket or the suburbanite quietly making out a check to the Sierra Club. The trend toward what is called green consumerism, represented in popular works such as *Fifty Easy Things You Can Do to Save the Planet* or *Shopping for a Green Planet*, is premised on the assumption that most environmental hazards are the result of irresponsible consumerism or shopping for a less-than-green planet. The rhetoric of such works suggests that we can all do something to save the environment, a generally acceptable statement, but—more problematically—that somehow the environmental conditions facing society today were similarly caused by "us." In this way, environmentalism is transformed into a critique that blames the shopper that asks for plastic bags at the supermarket but not the conglomerate that produces the bags, the supermarket chain that distributes them to local stores, the advertisers that place their names on the sides of these bags, or the corporate owners of all of the firms that fail to pose more environmentally sustainable alternatives for shoppers.

In all of this, the language of the environment fabricates an inclusiveness though its use of terms such as "we," "us," and "ours," all terms that obliterate the inequities in the environment. In these books, everyone suffers from environmental problems. "We" have to be concerned about "our" common future. "Our" children will inherit the world and it is for them that "we" have to recycle. The terms "we" and "us" cut against a politics that emphasizes the situatedness of the environment and the way it overlaps with social injustices and inequalities.

As a glance through any magazine will quickly reveal, the environment is a multi-billion dollar market. Even while criticizing the worst excesses of consumer culture, however, market-friendly environmentalism is wedded to it. Environmentally disposable diapers and biodegradable coffee filters all provide reassurances that these purchases will not harm the environment. A cleaner environment is just a purchase away.

For groups that work with the legacy of environmental hazards in the nation's poorest communities, the idea of the environment as a form of consumerism is seen as irrelevant at best. In the events surrounding the twentieth anniversary of Earth Day in April 1990, groups representing the poor and blue-collar communities in the environmental justice movement were critical of the rush to embrace environmental-friendly consumerism. Gary Cohen of the National Toxics Campaign was a leading critic of the 1990 Earth Day. "People were exhorted to buy environmentally friendly products, drive their cars less, recycle their bottles and newspapers, and plant trees wherever they could find the space" explains Cohen. Earth Day, writes the National Toxic Campaign's Cohen, "offered Americans an easy way out of our environmental crisis: just become better consumers" (Cohen 1990, 23). "People came to the party," concludes Cohen, "and went home feeling reassured they had done their 'bit' for the environment. They promised themselves they would recycle their newspapers and bottles and take shorter showers. The environment, everyone agreed, was in rough shape but somehow we could figure it out if we all worked together. The solution seemed fuzzy and the cause of the crisis seemed even fuzzier but Earth Day showed that we could all make a difference. Don't worry be happy" (Cohen 1990, 30). It lacked a focus on the socioeconomic changes and political struggles that would help the environment in poorer communities.

In the language used to refer to the environment, terms such as "we" and all of "us" clearly have implications for obscuring social relations of privilege. With this language of "we" and "us" and "our" common future, environmentalism is in danger of losing touch with the disparities highlighted by the environmental justice movement. As Gary Cohen writes, "we may all be passengers on Spaceship Earth but polluting corporations are clearly at the helm, with the rest of us choking on the fumes" (Cohen 1990, 22). Environmentalism is reduced to a decision to purchase recycled paper, while industry's disregard for the environment is unaddressed. To focus narrowly on consumption also has the effect of fissuring environmentalism itself away from the poor.

In responding to this, some of the groups in the environmental justice movement now recognize that the idiom of the environment as enlightened consumerism is leaving them without the tools of political struggle. "Environmentalists' message," explains the Environmental Careers Organization, a Boston-based group that works with many members of the environmental justice movement, "is geared to the middle and upper classes. 'We all must stop over-consuming' does not have much meaning to someone who can barely make ends meet, even if he or she is concerned about clean air or drinking water" (Environmental Careers Organization 1992). For all of these reasons, then, it is clear that the marketplace of environmental products is fraught with contradictions, as middle-class consumers associate an environmental conscience with a tumult of "green" products while millions of blue-collar and low-wage workers, living with the worst effects of environmental and hazardous wastes, remain detached from this culture of environmental lifestyles (as celebrated in the pages of magazines such as *The Utne Reader* and *Mother Jones*).

The Labor/Community Strategy Center in Los Angeles and other groups in the environmental justice movement are determined to challenge thinking about the environment in terms of narrow individualism. "There is a widespread view, encouraged by corporate propagandists," writes the Labor/Community Strategy Center, "that 'all of us' are responsible for the environmental crisis, and thus, 'all of us' should work together to clean it up. But while each one of us must take responsibility for our individual actions, it is a handful of powerful corporate executives who presently determine the life choices of the vast majority, and who must assume the lion's share of responsibility for the environmental dangers to public health and the threat to the planet's long term viability" (Mann and the Labor/Community Watchdog Organizing Committee 1991, 35). "Despite the rhetoric about how 'we all' are part of the problem," writes Eric Mann of the Labor/Community Strategy Center, "it is the corporations that determine industrial and agricultural production, and they must be brought under control" (Mann 1990a, 268). "We agree that individuals must take responsibility for making environmentally sound choices. But, for the most part, it is large corporations that manufacture the consumer products we purchase and that determine our choices through advertising, pricing, and other forms of power in the marketplace" (Mann and the Labor/Community Watchdog Organizing Committee 1991, 46). "It is misleading for us to talk about making environmentally sound 'choices' based on our individual consumption when it is Corporate America that must change its products in order for us to have any real options" (Mann and the Labor/Community Watchdog Organizing Committee 1991, 46). The Labor/Community Strategy Center was willing to challenge the narrow individualism of much of the environmental movement. Many more groups in the environmental justice movement and many members of the Gulf Coast Tenants Organization, the SouthWest Organizing Project and other groups also criticized the presumption that the contamination of the environment is something evenly felt with the same impact by all in society regardless of disparities in wealth, income, or political influence.

The unquestioned assumption that it is everybody's fault is linked to what Donna Lee King refers to as the "social reproduction" of an environmentalism that deflects political criticism of corporations or the private sector. "Responsibility is conveniently diffused. Corporate interests shielded yet again" (King 1994b, 50). This yields to solutions that more often than not do little to go far beyond the supermarket checkout counter or the office recycling bin. Focusing on the consumer on a limited budget who purchases nonbiodegradable products or, worse yet, accepts plastic bags at the supermarket checkout line, only leaves the politics of environmentalism vulnerable to the oft-repeated charge that environmentalists are an elite group that is unconcerned with the problems of working people and the poor.

In the 1990s, the issue of the environment threatens to dissolve into many different cultural fragments. The idea of environmental commonality—that there are environmental hazards shared with the same responsibility by all members of society—dissolves in the face of class fissures, economic polarization, and persistent racism.[2] The inclusiveness of environmental problems as they are typically understood ensures that the experiences of those groups most adversely impacted by environmental problems will all but be erased in the face of a future that is more likely to be characterized by widening class and socioeconomic inequalities, that will be mirrored by glaring inequalities in environmental despoliation and workplace health conditions, all amid a purportedly information-based, Third Wave, post-Cold War era.

RACE, POVERTY AND COMPETING NARRATIVES OF THE ENVIRONMENT

> Environmentalism is a deceptively unitary term that fails to register the blizzard of differences among people and groups who consider themselves part of it.
>
> —Marcy Darnovsky

The environment is connected with the lives of millions of people from all manner of backgrounds. The environment, writes Donna Haraway, "is constructed, constituted historically, not discovered naked in a fossil bed or a tropical forest. Nature is contested" (Haraway 1991). It is about chemical workers in Louisiana. It is about public housing tenants in Mississippi.

It is about poultry workers in North Carolina. It is about the single mother in Albuquerque worried about the violence from gangs in her neighborhood. It is about workers in an oil refinery in East Texas who know first-hand the health risks that their factory poses to nearby residents. The people in all of these places think about the environment in different ways. They might not use the language of environmental policy or relate to it through the issues that concern members of the Sierra Club or the Nature Conservancy, but the point is that the people who live in these places struggle to make sense of the environment on their own terms and in their own ways.

The struggle of the environmental justice movement and its fight to direct the attention of mainstream environmental groups to the poorest communities is altering the terrain of the environmental movement as well as promoting fascinating connections between issues. The environmental justice movement connects with the struggles of people in their communities, schools, workplaces and homes.

In an African-American community in South Georgia, an association made up largely of elderly women who purchase flowers for funerals of local residents is now leading the fight to draw attention to environmental conditions. In the San Francisco Bay area, Laotian fishing families are aligning with environmentalists to draw attention to the contamination of water in the bay. In the community of Columbia, Mississippi, residents are uniting behind Jesus People Against Pollution, a group that brings together the wellsprings of African-American religious faith with the struggle for the environment.

In the environmental justice movement, the work of these groups is drawn from distinctively different histories. African-American groups tend to ground their organizing in the history of the civil rights movement. Latino groups, in turn, point to a different history. The Southwest has a long tradition of struggle against encroachment on the land: Native American tribes resisting the destruction of their lands by mining, fishing, and lumber interests; Latino farmworkers who work with deadly pesticides without adequate protection. These are just some of the traditions of struggle drawn from by groups in the Southwest and throughout the country.

For too long, environmentalism has been beset by an assumption of the environment as a concern for more well-to-do groups. The environmental justice movement, however, is showing that groups from many different backgrounds have a stake in understanding and fighting to protect the environment. The environmental justice movement is contesting the very language used to describe the environment. It is also bringing to the fore a new history of environmentalism, one that simultaneously questions the history of mainstream environmentalism and raises new questions about the environment, race and social injustice.

It is imperative that the students of the 1990s learn about the diversity of cultural perspectives in their study of environmental problems as well as the international context of environmental problems and their overlap with the widening gap between the rich and poor of the global community.[3] Every day, new global actors in the international struggle for environmental justice emerge to show us the way to a rebuilding of political struggle that connects the environment with the circumstances of hardship and struggle experienced by untold millions around the globe.

The work of the environmental justice movement is clearing the way, so to speak, for diverse forms of cultural identity by coming to terms with the many different and varied definitions of the environment. The crucial work of the environmental justice movement is to define language itself as the site of political struggle, where individuals can fight to reclaim their own definition of what constitutes the environment and struggle to reconfigure the scientific, legal, and technical definitions of the environment in a more accessible language. Language itself is a site of political struggle, a part of activism no less important than the protest at a chemical facility, an information table at an Earth Day rally, or the demonstrations at a hazardous waste site by communities that for too long have not been a part of the postwar environmental movement.

NOTES

1. As Fields writes, "to put it another way, part of what human beings understand by the word 'nature' is the sense of inevitability that gradually becomes attached to a predictable, repetitive social routine, 'custom, so immemorial that it looks like nature,' as Nathaniel Hawthorne wrote" (Fields 1990, 106).

2. Marcy Darnovsky writes that if "environmental degradations become predicaments that 'we' suffer and will solve together rather than disproportionately affecting different groups of people, they become problems amenable to technical fixes rather than symptoms of systemic political and economic injustices" (Darnovsky 1992).

3. The Centre for Science and Environment in Kailash Colony, New Delhi is a leading group working on environmental problems in the context of economic inequities and international politics. At the United Nations Conference on Environment and Development, it, along with a thousand other organizations, had an historic opportunity to formulate a response to the environmental crisis in an assembly that recognized social, economic, and political divisions in the international community.

Bibliography

Acuna, Rodolfo. 1972. *Occupied America: The Chicano's Struggle Toward Liberation*. San Francisco: Canfield Press.

Albuquerque Journal. 1970. "Earth Day Will Tune in to Turn off Pollution." *Albuquerque Journal* (April 22), D1.

Alston, Dana. 1990. *Taking Back Our Lives: A Report to the Panos Institute on Environment, Community Development and Race in the United States*. Washington, D.C.: The Panos Institute.

Anthony, Carl. 1990. "Why African Americans Should Be Environmentalists." *Race, Poverty and the Environment: A Newsletter for Social and Environmental Justice* 1, no. 1 (April), 5–6.

Beasley, Jr., Conger. 1990. "Of Pollution and Poverty: Keeping Watch in Cancer Alley." *Buzzworm: The Environmental Journal* 2, no. 4 (July/August), 38–45.

Benford, Robert D. 1993a. "'You Could Be the Hundredth Monkey': Collective Action Frames and Vocabularies of Motive within the Nuclear Disarmament Movement." *The Sociological Quarterly* 34, no. 2 (May), 195–216.

Benford, Robert D. 1993b. "Frame Disputes within the Nuclear Disarmament Movement." *Social Forces* 71, no. 3 (March), 677–701.

Benford, Robert D. and Louis A. Zurcher. 1990. "Instrumental and Symbolic Competition Among Peace Movement Organizations." In Sam Marullo and John Lofland (eds.) *Peace Action in the Eighties: Social Science Perspectives*. New Brunswick: Rutgers University Press, 125–139.

Benford, Robert D. and Scott A. Hunt. 1992. "Dramaturgy and Social Movements: The Social Construction and Communication of Power." *Sociological Inquiry* 62, no. 1 (February), 36–55.

Bennett, Bonnie. 1991. "Mike Guerrero: A Community Organizer's Profile." *Voces Unidas* 1, no. 3 (Fall) [SouthWest Organizing Project].

Berry, Jason. 1984. "The Poisoning of Louisiana." *Southern Exposure* 12, no. 2 (March/April), 16–23.

Bloch, Robin and Roger Keil. 1991. "Planning for a Fragrant Future: Air Pollution Control, Restructuring, and Popular Alternatives in Los Angeles." *Capitalism, Nature, Socialism* 2, no. 1 (February), 44–65.

Bowermaster, Jon. 1993. "A Town Called Morrisonville: In Louisiana's Cancer Alley, Dow Chemical and Other Companies Are Buying Out Neighboring Communities." *Audubon* 95, no. 4 (July), 42–50.

Braden, Anne. 1993. "Environmental Justice and Social Justice: Southern Organizing Heralds New Movement." *Labor Research Review*, no. 20 (January), 63–69.

Bryant, Bunyan. 1993. "Racism: A Toxic Waste." *Against the Current* 7, no. 6 (January/February), 10–12.

Bryant, Bunyan. 1995. (ed.) *Environmental Justice: Issues, Policies and Solutions*. Washington: Island Press.

Bryant, Bunyan and Paul Mohai. 1992. (eds.) *Race and the Incidence of Environmental Hazards: A Time for Discourse*. Boulder: Westview Press.

Bryant, Pat. 1982. "Turning Toward Life." *Southern Exposure* 10, no. 6 (November/December), 28–31.

Bryant, Pat. 1989. "Toxics and Racial Justice." *Social Policy* 20, no. 1 (Summer), 48–52.

Bryant, Pat. 1990a. *Annual Report to the Membership and Board of the Gulf Coast Tenants Leadership Project and the Gulf Coast Tenants Organization* 1990, New Orleans.

Bryant, Pat. 1990b. "A Lily White Achilles Heel." *Environmental Action* 21, no. 4 (January/February), 28–29.

Bryant, Pat. 1992a. "A View from Down Below." In Charles Lee (ed.) *Proceedings of the First National People of Color Environmental Leadership Summit*. New York: United Church of Christ Commission for Racial Justice, 83–85.

Bryant, Pat. 1992b. "Milestone of Hope and Solidarity." *Crossroads/Forward Motion* 11, no. 2 (April), 2–6.

Bryant, Pat. 1993a. *Fertile Ground Between Environmentalists and People of Color*. New Orleans: Gulf Coast Tenants Organization.

Bryant, Pat. 1993b. "Tenants and Toxics." *Southern Exposure* 21, no. 4 (December), 27–29.

Bryant, Pat. 1993c. "Hon Rule EPA Hearing." *Gulf Coast Tenant's Voice* 11, no. 2 (April).

Bullard, Robert D. 1990. *Dumping in Dixie: Race, Class and Environmental Quality*. Boulder: Westview Press.

Bullard, Robert D. 1993a. "Anatomy of Environmental Racism and the Environmental Justice Movement." In Robert D. Bullard (ed.) *Con-*

fronting Environmental Racism: Voices from the Grassroots. Boston: South End Press, 15–39.

Bullard, Robert D. 1993b. (ed.) *Unequal Protection: Environmental Justice and Communities of Color.* San Francisco: Sierra Club Books.

Bullard, Robert D. 1994a. *Dumping in Dixie: Race, Class and Environmental Quality.* 2nd ed. Boulder: Westview Press.

Bullard, Robert D. 1994b. *People of Color Environmental Groups, 1994–1995 Directory.* Flint, Michigan: Charles Stewart Mott Foundation.

Bullard, Robert D. and Beverly Hendrix Wright. 1987. "Blacks and the Environment." *Humboldt Journal of Social Relations* 14, nos. 1–2 (Fall/Winter), 165–184.

Bullard, Robert D. and Beverly H. Wright. 1991. "The Quest for Environmental Equity: Mobilizing the African-American Community for Social Change." *Society and Natural Resources* 3, no. 4 (October/December), 301–311.

Capek, Stella M. 1993. "The 'Environmental Justice' Frame: A Conceptual Discussion and an Application." *Social Problems* 40, no. 1 (February), 5–24.

Carson, Rachel. 1962. *Silent Spring.* Boston: Houghton Mifflin.

Chavis, Jr., Benjamin F. 1994. "Preface." In Robert D. Bullard (ed.) *Unequal Protection: Environmental Justice and Communities of Color.* San Francisco: Sierra Club Books, xi–xii.

Chavkin, Wendy. 1984. (ed.) *Double Exposure: Women's Health Hazards on the Job and at Home.* New York: Monthly Review Press.

Cheakalos, Christina. 1993. "A Thin Green Line: Louisiana Town Fights a Chemical Giant." *The Atlanta Constitution* (February 23), B4–B5.

Citizen's Clearinghouse for Hazardous Wastes. 1986. *Five Years of Progress, 1981–1986: A Commemorative History and Call to Action, the First National Grassroots Convention on Toxics.* Arlington, Virginia: Citizen's Clearinghouse for Hazardous Wastes [May].

Citizen's Clearinghouse for Hazardous Wastes. 1990. "People United for Environmental Justice." *Everybody's Backyard* 8, no. 1 (January/February), 2.

Citizen's Clearinghouse for Hazardous Wastes. 1993. *Ten Years of Triumph, Turning the Tide: Ten Year Anniversary of the Grassroots Movement for Environmental Justice.* Falls Church, Virginia: Citizen's Clearinghouse for Hazardous Wastes [May].

Cobb, James C. 1982. *The Selling of the South: The Southern Crusade for Industrial Development, 1936–1980.* Baton Rouge: Louisiana State University Press.

Cockburn, Alexander and Jeffrey St. Clair. 1994. "Death and Life for America's Greens." *The Nation* (December 19), 760–765.

Cohen, Gary. 1990. "It's Too Easy Being Green: Earth Day 1990." *Social Policy* 21, no. 1 (Summer), 20–31.

Cohen, Michael P. 1988. *A History of the Sierra Club, 1892–1970*. San Francisco: Sierra Club Books.

Cole, Luke W. 1992. "Empowerment as the Key to Environmental Protection: The Need for Environmental Poverty Law." *Ecology Law Quarterly* 19, no. 4 (March), 619–683.

Commoner, Barry. 1990. *Making Peace with the Planet*. New York: Pantheon Books.

Commoner, Barry. 1992. "Yearning to Breathe Free." *The Nation* (February 24), 242–243.

Costner, Pat and Joe Thornton. 1989. *We All Live Downstream: The Mississippi River and the National Toxics Crisis*. Washington: Greenpeace.

Cotgrove, Steven and Andrew Duff. 1980. "Environmentalism, Middle Class Radicalism and Politics." *Sociological Review* 28, no. 2 (May), 333–351.

Couto, Richard A. 1993. "Narrative, Free Space and Political Leadership in Social Movements." *The Journal of Politics* 55, no. 1 (February), 57–79.

Darnovsky, Marcy. 1992. "Stories Less Told: Histories of U.S. Environmentalism." *Socialist Review* 22, no. 4 (October/December), 11–54.

Davis, Mike. 1992. "The L.A. Inferno." *Socialist Review* 22, no. 1 (January/March), 57–80.

Davis, Mike. 1993. "Dead West: Ecocide in Marlboro Country." *New Left Review* 200 (July/August), 49–73.

Di Chiro, Giovanna. 1992. "Defining Environmental Justice: Women's Voices and Grassroots Politics." *Socialist Review* 22, no. 4 (October/December), 93–130.

Dowie, Mark. 1995. *Losing Ground: American Environmentalism at the Close of the Twentieth Century*. Cambridge, Massachusetts: The MIT Press.

Dunlap, Riley E. and Kent D. Van Liere. 1978. "The 'New Environmental Paradigm': A Proposed Measuring Instrument and Preliminary Results." *The Journal of Environmental Education* 9, 10–19.

Eagleton, Terry. 1983. *Literary Theory: An Introduction*. Minneapolis: The University of Minnesota Press.

Earthworks Group. *Fifty Simple Things You Can Do to Save the Earth*. Berkeley: Earthworks Group.

Entman, Robert M. 1993. "Framing, Toward Clarification of a Fractured Paradigm." *The Journal of Communication* 43, no. 4 (Autumn), 51–58.

Environmental Action. 1970. (ed.) *Earth Day: The Beginning*. New York: Arno Press.

Environmental Careers Organization. 1992. *Beyond the Green: Redefining and Diversifying the Environmental Movement*. Boston: Environmental Careers Organization.

Epstein, Barbara. 1991. *Political Protest and Cultural Revolution: Nonviolent Direct Action in the 1970s and 1980s*. Berkeley: University of California Press.

Epstein, Robin. 1993. "Making the Fat Cats Chip In." *In These Times* 17, no. 6 (February 8), 23–25.

Evans, Larry. 1991. "Louisiana Labor and Neighbor Alliance." *The Louisiana Labor/Neighbor Journal* (Spring), 5.

Faber, Daniel and James O'Connor. 1989. "The Struggle for Nature: Environmental Crises and the Crisis of Environmentalism in the United States." *Capitalism, Nature, Socialism* 2 (Summer), 12–39.

Factor, Diane and Joe Uehlein. 1991. "Organizing for Safe Work in a Safe World." *Labor Research Review* 16 (Spring), 1–13.

Fantasia, Rick. 1988. *Cultures of Solidarity: Consciousness, Action and Contemporary American Workers*. Berkeley: University of California Press.

Fields, Barbara Jeanne. 1990. "Slavery, Race and Ideology in the United States," *New Left Review* 181 (May), 95–118.

Florini, Karen L., George D. Krumbhaar and Ellen K. Silbergeld. 1990. *Legacy of Lead: America's Continuing Epidemic of Childhood Lead Poisoning*. Washington, D.C. : Environmental Defense Fund.

Forrest, Suzanne. 1989. *The Preservation of the Village: New Mexico's Hispanics and the New Deal*. Albuquerque: University of New Mexico Press.

Fox, Stephen. 1985. *The American Conservation Movement: John Muir and His Legacy*. Madison: University of Wisconsin Press.

Freedman, Carl. 1992. "Louisiana Duce: Toward a Systematic Analysis of Postmodern Fascism in America." *Rethinking Marxism* 5, no. 1 (Summer), 19–31.

Freudenberg, Nicholas. 1984. *Not in Our Backyards: Community Action for Health and the Environment*. New York: Monthly Review Press.

Friedman, Debra and Doug McAdam. 1992. "Collective Identity and Activism: Networks, Choices and the Life of a Social Movement." In Aldon D. Morris and Carol McClurg Mueller (eds.) *Frontiers in Social Movement Theory*. New Haven: Yale University Press, 156–173.

Gauna, Jeanne. 1991. *Unsafe for Women, Children and Other Living Things*. Albuquerque: SouthWest Organizing Project.

Gauna, Jeanne. 1992. "Exposing New Mexico's Deadly Clean Industry." *Voces Unidas* 2, no. 2 (December), 1.

Gauna, Jeanne. 1994. "Environmental and Economic Justice in New Mexico." *The Workbook* 19, no. 2 (Summer), 79–80.

Geiser, Ken. 1983. "Toxic Times and Class Politics." *Radical America* 17, no. 2/3 (March), 38–50.

Geiser, Ken. 1991. "Seeking Clean and Common Ground." *Dollars and Sense* 164 (March), 16–18.

George, Larry N. 1982. "Love Canal and the Politics of Corporate Terrorism." *Socialist Review* 12, no. 6 (November/December), 9–38.

Gismondi, Michael and Mary Richardson. 1994. "Discourse and Power in Environmental Politics: Public Hearings on a Bleached Kraft Pulp Mill in Alberta, Canada." In Martin O'Connor (ed.) *Is Capitalism Sustainable? Political Economy and the Politics of Ecology*. New York: Guilford Press, 232–252.

Goffman, Erving. 1974. *Frame Analysis: An Essay on the Organization of Experience*. Cambridge, Massachusetts: Harvard University Press.

Goldman, Benjamin A. 1991. *The Truth About Where You Live: An Atlas for Action On Toxins and Mortality*. New York: Times Books.

Gottlieb, Robert. 1992. "A Question of Class: The Workplace Experience." *Socialist Review* 22, no. 4 (October/December), 131–165.

Gottlieb, Robert. 1993. *Forcing the Spring: The Transformation of the American Environmental Movement*. Washington, D.C.: Island Press.

Gozan, Julie. 1993. "Building Bridges: Labor/Community Alliances Grow in Wake of Longest Labor Lockout." *OCAW Reporter* 49, nos. 7–8 (September/October), 10–11.

Gremillion, Sr., Esnard. 1987. *Letter to Louisiana Environmental Action Network*. Baton Rouge: Oil, Chemical and Atomic Workers (October 12).

Gremillion, Sr., Esnard. 1991. "OCAW Red Stick Local 4-620's Capsule History." *The Louisiana Labor/Neighbor Journal* (Spring).

Grossman, Karl. 1991. "Environmental Racism." *The Crisis* 98, no. 4 (April), 14–17.

Grossman, Richard. 1985. "Environmentalists and the Labor Movement." *Socialist Review* 82/83 (July/October), 63–87.

Guerrero, Michael. 1992. "Strategy Group Workshop: Grassroots Empowerment." In Charles Lee (ed.) *Proceedings of the First National People of Color Environmental Leadership Summit*. New York: United Church of Christ Commission for Racial Justice, 149–150.

Guerrero, Michael and Louis Head. 1990a. "The Environment: Redefining the Issue." In Dana Alston (ed.) *We Speak for Ourselves: Social Justice, Race and Environment*. Washington, D.C.: The Panos Institute (December), 11.

Guerrero, Michael and Louis Head. 1990b. "Informed Collective Action: A Powerful Weapon." In Dana Alston (ed.) *We Speak for Ourselves: Social Justice, Race and Environment*. Washington: The Panos Institute (December), 24–25.

Gulf Coast Tenants Leadership Development Project. 1989. *Developing Tenants and Community Leaders to Build Long-Range Coalitions for Peace and Justice: An Education Program for Decent Housing, a Safe Environment and Peace* [New Orleans].

Gulf Coast Tenants Organization. 1989. *No Place to Live: Build Houses, Not Bombs*. New Orleans: Gulf Coast Tenants Organization [January].

Gulf Coast Tenants Organization. 1990a. *Louisiana Great Toxics March Principles of Unity*. Baton Rouge: Gulf Coast Tenants Organization [March].

Gulf Coast Tenants Organization. 1990b. *Come Join the Second Great Louisiana March Against Poisons*. New Orleans: Gulf Coast Tenants Organization [April].

Gulf Coast Tenants Organization. 1990c. "Caravan to Fight Lost Constitutional Rights, Jobs, Housing and Clean Environment." *Gulf Coast Tenants Voice* 1, no. 1, August/September 1990, 1.

Gulf Coast Tenants Organization. 1991. *Racial Justice, Peace, Housing, Jobs, Environment: The Program of the Gulf Coast Tenants Organization*. Gulf Coast Tenants Organization: New Orleans [flyer].

Gulf Coast Tenants Organization. 1992. *A Short History of the Gulf Coast Tenants Organization*. New Orleans: Gulf Coast Tenants Organization.

Gulf Coast Tenants Organization. 1993a. *Southern Action Manifesto for Environmental and Economic Justice*. New Orleans: Gulf Coast Tenants Organization [September].

Gulf Coast Tenants Organization. 1993b. *African-American Environmental Education Program, Prepared for Lyons, Louisiana Community Leaders*. New Orleans: Gulf Coast Tenants Organization.

Gulf Coast Tenants Organization. 1993c. *Digest of the Southern Action Manifesto for Environmental and Economic Justice: End the Poisoning of Our Communities, Create Safe and Healthy Workplaces, Provide Jobs, Education, Housing, Health Care and a Clean Environment. Gulf Coast Tenants Voice* 6, no. 1 (December), 11.

Gulf Coast Tenants Organization. 1993d. *Racism Destroys Environment*. New Orleans: Gulf Coast Tenants Organization [September].

Gulf Coast Tenants Organization. 1994. *Gulf Coast Tenants Organization*. Baton Rouge: Gulf Coast Tenants Organization.

Gulf Coast Tenants Organization and Southern Organizing Committee for Economic and Social Justice. 1993. *In Communities Across the South*. Birmingham, Alabama: Southern Organizing Committee for Economic and Social Justice.

Hager, George. 1999. "Study Reports Record U.S. Income Gap." *The Washington Post* (September 5), A8.

Hall, Bob. 1988. (ed.) *Environmental Politics: Lessons from the Grassroots*. Durham: Institute for Southern Studies.

Hall, Bob and Mary Lee Kerr. 1991. *The 1991–1992 Green Index: A State-by-State Guide to the Nation's Environmental Health*. Durham: Institute for Southern Studies and Island Press.

Hall, Thomas D. 1989. *Social Change in the Southwest, 1350–1880*. Lawrence: University Press of Kansas.

Hamilton, James T. 1993. "Politics and Social Costs: Estimating the Impact of Collective Action on Hazardous Waste Facilities." *Rand Journal of Economics* 24, no.1 (Summer), 101–125.

Hamilton, James T. 1995. "Testing for Environmental Racism: Prejudice, Profits, Political Power?" *The Journal of Policy Analysis and Management* 14, no. 1 (Winter), 107–132.

Haraway, Donna J. 1991. *Simians, Cyborgs and Women: The Reinvention of Nature*. New York: Routledge.

Hare, Nathan. 1970. "Black Ecology." *The Black Scholar* 1, no. 6 (April), 2–8.

Harvey, David. 1993. "The Nature of Environment: Dialectics of Social and Environmental Change." In Ralph Miliband and Leo Panitch (eds.) *Real Problems, False Solutions: Socialist Register 1993*. London: The Merlin Press, 1–51.

Harvey, David. 1996. *Justice, Nature and the Geography of Difference*. London: Blackwell Publishers.

Hays, Samuel P. 1987. *Beauty, Health and Permanence: Environmental Politics in the United States, 1955–1985*. Cambridge: Cambridge University Press.

Head, Louis and Michael Guerrero. 1991. *Fighting Environmental Racism*. Albuquerque: SouthWest Organizing Project.

Helvarg, David. 1994. *The War Against the Greens: The Wise Use Movement, the New Right and Anti-Environmental Violence*. San Francisco: Sierra Club Books.

Hershey, Marjorie Randon and David B. Hill. 1977. "Is Pollution 'A White Thing'? Racial Differences in Preadults' Attitudes." *Public Opinion Quarterly* 41, no. 4 (Winter), 439–458.

Hooks, Benjamin L. "Forward." *The Crisis* 98, no. 4 (April), 4 [National Association for the Advancement of Colored People].

Houck, Oliver A. 1986. "This Side of Heresy: Conditioning Louisiana's Ten-Year Industrial Tax Exemption Upon Compliance with Environmental Laws." *Tulane Law Review* 61, no. 2 (December), 289–377.

Hunt, Scott A. and Robert D. Benford. 1994. "Identity Talk in the Peace and Justice Movement." *The Journal of Contemporary Ethnography* 22, no. 4 (January), 488–517.

Hunt, Scott A., Robert D. Benford and David A. Snow. 1994. "Identity Fields: Framing Processes and the Social Construction of Movement Identities." In Enrique Larana, Hank Johnston and Joseph R. Gusfield (eds.) *New Social Movements: From Ideology to Identity*. Philadelphia: Temple University Press, 185–208.

Hurley, Andrew. 1992. "Challenging Corporate Polluters: Race, Class and Environmental Politics in Gary, Indiana Since 1945." *Indiana Magazine of History* 88, no. 4 (December), 273–302.

Hurley, Andrew. 1995. *Environmental Inequalities: Class, Race and Industrial Pollution in Gary, Indiana, 1945–1980.* Chapel Hill: The University of North Carolina Press.

Inglehart, Ronald. 1977. *The Silent Revolution: Changing Values and Political Styles Among Western Publics.* Princeton: Princeton University Press.

Inglehart, Ronald. 1990. *Culture Shift in Advanced Industrial Society.* Princeton, New Jersey: Princeton University Press.

Jamison, Andrew, Ron Eyerman and Jacqueline Cramer. 1990. *The Making of the New Environmental Consciousness: A Comparative Study of the Environmental Movements in Sweden, Denmark and the Netherlands.* Edinburgh: Edinburgh University Press.

Jenkins, J. Craig. 1983. "Resource Mobilization Theory and the Study of Social Movements." In Ralph H. Turner and James F. Short (eds.) *Annual Review of Sociology.* Palo Alto, California: Annual Reviews Incorporated, 527–553.

Johnston, Hank. 1991. *Tales of Nationalism: Catalonia, 1939–1979.* New Brunswick: Rutgers University Press.

Jones, Robert Emmet and Lewis F. Carter. 1994. "Concern for the Environment Among Black Americans: An Assessment of Common Assumptions." *Social Science Quarterly* 75, no. 3 (September), 560–579.

Jones, Tara. 1988. *Corporate Killing: Bophals Will Happen.* London: Free Association Books.

Kallick, David. 1990. "The Struggle for Community: Race, Class and the Environment—An Interview with Cynthia Hamilton." *Social Policy* 21, no. 2 (Fall), 18–25.

Kay, Jane. 1993. "California's Endangered Communities of Color." In Robert D. Bullard (ed.) *Unequal Protection: Environmental Justice and Communities of Color.* San Francisco: Sierra Club Books, 155–188.

Kazis, Richard and Richard L. Grossman. 1982. *Fear at Work: Job Blackmail, Labor and the Environment.* New York: Pilgrim Press.

Kerr, Mary Lee and Charles Lee. 1993. "From Conquistadors to Coalitions: After Centuries of Environmental Racism, People of Color Are Forging a New Movement for Environmental Justice." *Southern Exposure* 21, no. 4 (Winter), 9–19.

King, Donna Lee. 1994a. "Captain Planet and the Planeteers: Kids, Environmental Crisis and Competing Narratives of the New World Order." *The Sociological Quarterly* 35, no. 1 (January), 103–120.

King, Donna Lee. 1994b. *Doing Their Share to Save the Planet: Children and Environmental Crisis.* New Brunswick, New Jersey: Rutgers University Press.

Krieger, Nancy, Diane L. Rowley, Allen A. Herman, Byllye Avery and Mona T. Phillips. 1993. "Racism, Sexism and Social Class: Implications for Studies of Health, Disease and Well-Being." In Diane Rowley and Heather Tosteson (eds.) *Racial Differences in Pre-Term Delivery: De-*

veloping a New Research Paradigm. New York: Oxford University Press, 82–122.

Labalme, Jenny. 1987. *A Road to Walk: A Struggle for Environmental Justice.* Durham, North Carolina: Regulator Bookshop.

Labor/Community Strategy Center. 1990. *The Labor/Community Watchdog: A Campaign for Workers, Communities and People of Color to Have a Direct Impact on the Air Quality of Los Angeles and the Policies of the South Coast Air Quality Management District.* Los Angeles: Labor/Community Strategy Center.

Labor/Community Strategy Center. 1993a. *Reconstructing Los Angeles from the Bottom Up: A Long-Term Strategy for Workers, Low-Income People, and People of Color to Create an Alternative Vision of Urban Development.* Los Angeles: Labor/Community Strategy Center.

Labor/Community Strategy Center. 1993b. *L.A.'s Lethal Air: It's Time for a Change.* Wilmington: Labor/Community Strategy Center [flyer].

Labor/Community Strategy Center. 1994a. *An Open Letter to the Workers at Texaco's Wilmington Refinery from the Labor/Community Watchdog.* Los Angeles: Labor/Community Strategy Center [February].

Labor/Community Strategy Center. 1994b. *Boycott Texaco.* Los Angeles: Labor/Community Strategy Center [flyer].

Lazarovici, Lauren. 1991. "Air Battles." *L.A. Weekly* (December 6), 1.

Lee, Charles. 1992a. "Toxic Waste and Race in the United States." In Bunyan Bryant and Paul Mohai (eds.) *Race and the Incidence of Environmental Hazards: A Time for Discourse.* Boulder: Westview Press, 10–27.

Lee, Charles. 1992b. "Introduction." In *Proceedings of the First National People of Color Environmental Leadership Summit.* New York: United Church of Christ Commission for Racial Justice, v–xi.

Lee, Charles. 1993a. "Beyond Toxic Wastes and Race." In Robert D. Bullard (ed.) *Confronting Environmental Racism: Voices from the Grassroots.* Boston: South End Press, 41–52.

Lee, Charles. 1993b. "From Los Angeles, East St. Louis and Matamoros: Developing Working Definitions of Urban Environmental Justice." *Earth Island Journal* 8, no. 4 (Fall), 39–42.

Leonard, Richard and Zack Nauth. 1990. "Beating BASF: OCAW Busts Union-Buster." *Labor Research Review* 9, no. 2 (Fall), 35–49.

Lewis, Sanford, Brian Keating and Dick Russell. 1992. *Inconclusive by Design: Waste, Fraud and Abuse in Federal Environmental Health Research.* Boston: Environmental Health Network and National Toxics Campaign Fund.

Louisiana Advisory Committee to the United States Commission on Civil Rights. 1993. *The Battle for Environmental Justice in Louisiana: Government, Industry and the People.* Louisiana Advisory Committee [September].

Louisiana Coalition for Tax Justice. 1992. *The Great Louisiana Tax Giveaway: A Decade Of Corporate Welfare, 1980–1989*. Baton Rouge: Louisiana Coalition for Tax Justice.

Louisiana Toxics Project. 1988. The Great Louisiana Toxics March. New Orleans: Louisiana Toxics Project (October 27).

Lowe, Philip D. and Wolfgang Rudig. 1986. "Political Ecology and the Social Sciences: The State of the Art." *British Journal of Political Science* 16, no. 4 (October), 513–550.

Lundy, Reverend George F. and William Temmink. 1987. "A Battle on the Mississippi." *Blueprint for Justice* 25, no. 8 (April) [Loyola University, New Orleans], 1–8.

MacLean, Alair. 1993. "Bigotry and Poison." *The Progressive* 57, no. 1 (January), 14.

Madison, Isaiah. 1993. "To Forge a Movement: How Can We Build a Broad Movement for Environmental Justice? Seventeen Grassroots Organizers Share Their Ideas and Experiences." *Southern Exposure* 21, no. 4 (December), 30–32.

Madison, Isaiah, Vernice Miller and Charles Lee. 1992. "The Principles of Environmental Justice: Formation and Meaning." In Charles Lee (ed.) *Proceedings of the First National People of Color Environmental Leadership Summit*. New York: United Church of Christ Commission for Racial Justice, 49–55.

Malek-Wiley, Darryl. 1988. "The Great Louisiana Toxic March." *Blueprint for Social Justice* 15, no. 2 (October), 1–7 [Institute of Human Relations, Loyola University, New Orleans].

Mann, Eric. 1987. *Taking On General Motors: A Case Study of the UAW Campaign to Keep GM Van Nuys Open*. Los Angeles: Center for Labor Research and Education, Institute of Industrial Relations, University of California.

Mann, Eric. 1988. "Keeping GM Van Nuys Open." In Mike Davis and Michael Sprinker (eds.) *Reshaping the U.S. Left: Popular Struggles in the 1980s*. London: Verso Press, 183–209.

Mann, Eric. 1990a. "Labor/Community Coalitions as a Tactic for Labor Insurgency." In Jeremy Brecher and Tim Costello (eds.) *Building Bridges: The Emerging Grassroots Coalition of Labor and Community*. New York: Monthly Review Press, 113–132.

Mann, Eric. 1990b. "Environmentalism in the Corporate Climate." *Tikkun* 5, no. 2 (March/April), 60–65.

Mann, Eric. 1990c. "Lighting a Spark: L.A.'s Smogbusters." *The Nation* (September 17), 267–272.

Mann, Eric. 1992. "Creating a Labor Strategy." *Environmental Action* 24, no. 1 (Spring), 16–17.

Mann, Eric. 1993. "Labor's Environmental Agenda in the New Corporate Climate." In Richard Hofrichter (ed.) *Toxic Struggles: The Theory and*

Practice of Environmental Justice. Philadelphia: New Society Publishers, 179–185.

Mann, Eric and the Labor/Community Watchdog Organizing Committee. 1991. *L.A.'s Lethal Air: New Strategies for Policy, Organizing and Action*. Van Nuys: Labor/Community Strategy Center.

Marsh, Alan. 1975. "The 'Silent Revolution,' Value Priorities and the Quality of Life in Britain." *American Political Science Review* 69, no. 1 (March), 21–30.

Martinez, Elizabeth. 1991a. *When People of Color Are an Endangered Species*. Albuquerque, New Mexico: SouthWest Organizing Project.

Martinez, Elizabeth. 1991b. *500 Anos del Pueblo Chicano/500 Years of Chicano History*. Albuquerque: SouthWest Organizing Project.

Martinez, Elizabeth and Louis Head. 1992. "Media White-Out Environmental Racism." *Extra!* 5, no. 5 (July/August), 29–31 [Fairness and Accuracy in Reporting, New York].

Martinez-Alier, Joan. 1998. "Environmental Justice (Local and Global)." In Fredric Jameson and Masao Miyoshi (eds.) *The Cultures of Globalization*. Durham, North Carolina: Duke University Press, 312–326.

McAdam, Doug. 1982. *Political Process and the Development of Black Insurgency, 1930–1970*. Chicago: University of Chicago Press.

McAdam, Doug. 1994. "Culture and Social Movements." In Enrique Larana, Hank Johnston and Joseph R. Gusfield (eds.) *New Social Movements: From Ideology to Identity*. Philadelphia: Temple University Press, 36–57.

McCarthy, John D. and Mayer N. Zald. 1979. "Resource Mobilization and Social Movements: A Partial Theory." *American Journal of Sociology* 82, no. 6 (May), 1212–1241.

McLure, Laura. 1990. "Labor, Ecologists Grapple with Tensions." *The Guardian* 42, no. 6 (April), 4.

Melucci, Alberto. 1989. *Nomads of the Present: Social Movements and Individual Needs in Contemporary Society*. Edited by John Keane and Paul Mier. Philadelphia: Temple University Press.

Milbrath, Lester W. 1984. *Environmentalists: Vanguard for a New Society*. Albany: State University of New York Press.

Miller, Richard. 1991. "A Sketch of OCAW's History in Occupational Health Regulation." *Louisiana Labor/Neighbor Journal* (Spring), 62–65.

Mohai, Paul. 1985. "Public Concern and Elite Involvement in Environmental Conservation Issues." *Social Science Quarterly* 66, no. 4 (December), 820–838.

Mohai, Paul. 1990. "Black Environmentalism." *Social Science Quarterly* 71, no. 4 (December), 744–765.

Moody, Kim. 1988. *An Injury to All: The Decline of American Unionism*. London: Verso.

Moody, Kim. 1990. "Union Gets Contract at BASF: Corporate Campaign Credited with Ending Five Year Lock-Out." *Labor Notes* 131 (February), 1–2.

Mooney, Patrick H. and Scott A. Hunt. 1996. "A Repertoire of Interpretations: Master Frames and Ideological Continuity in U.S. Agrarian Mobilization." *The Sociological Quarterly* 37, no. 1 (Winter), 177–197.

Moore, Richard. 1991a. *Toxics, Race and Class: The Poisoning of Communities*. Albuquerque, New Mexico: SouthWest Organizing Project.

Moore, Richard. 1991b. "Social Justice and the Environmental Movement." In Bunyan Bryant and Paul Mohai (eds.) *Environmental Racism: Issues and Dilemmas, a University of Michigan Symposium on Martin Luther King*. Ann Arbor: University of Michigan Office of Minority Affairs, 61–65.

Moore, Richard. 1994. "The Network for Environmental and Economic Justice in the Southwest: An Interview with Richard Moore." *Capitalism, Nature, Socialism* 5, no. 1 (March), 21–54.

Moore, Richard and Louis Head. 1993. "Acknowledging the Past, Confronting the Present: Environmental Justice in the 1990s." In Richard Hofrichter (ed.) *Toxic Struggles: The Theory and Practice of Environmental Justice*. Philadelphia: New Society Publishers, 118–127.

Morris, Aldon D. 1984. *The Origins of the Civil Rights Movement: Black Communities Organizing for Change*. New York: Free Press.

Morris, Aldon D. 1992. "Political Consciousness and Collective Action." In Aldon D. Morris and Carol McClurg Mueller (eds.) *Frontiers in Social Movement Theory*. New Haven, Connecticut: Yale University Press, 351–373.

Morris, Aldon D., Shirley J. Hatchett and Ronald E. Brown. 1989. "The Civil Rights Movement and Black Political Socialization." In Roberta S. Sigel (ed.) *Political Learning in Adulthood: A Sourcebook of Theory and Research*. Chicago: University of Chicago Press, 272–305.

Nabokov, Peter. 1969. *Tijerina and the Courthouse Raid*. Albuquerque: University of New Mexico Press.

National Association for the Advancement of Colored People and United Church of Christ Commission for Racial Justice. 1994. *Toxic Wastes and Race Revisited: An Update of the 1987 Report on the Racial and Socioeconomic Characteristics of Communities with Hazardous Waste Sites*. (Washington: Center for Policy Alternatives) [August].

Nauth, Zack. 1992. "Push Comes to Shove: Chemical Workers Break the BASF Lock-Out." In Sanford J. Lewis, *The Good Neighbor Handbook: A Community-Based Strategy for Sustainable Industry*. Boston: National Toxics Campaign Fund, 24–31.

Novak, Shonda. 1993. "Activists Blast Intel Tax Breaks." *The Albuquerque Tribune* (April 2), B10.

Novotny, Patrick. 1997. "The Peace and Justice Movement in the South in the 1980s." *Peace and Change* 22, no. 2 (April), 154–174.

O'Connor, James. 1995. "Three Ways to Think About the Ecological History of Monterey Bay," *Capitalism, Nature, Socialism* 6 (June), 21–47.

Oil, Chemical and Atomic Workers International Union. 1992. *Understanding the Conflict Between Jobs and the Environment: A Preliminary Discussion of the Superfund for Workers Concept.* Denver: Oil, Chemical and Atomic Workers International Union.

Oil, Chemical and Atomic Workers International Union. 1993. "Organizing as a Way of Life in OCAW: Why We Must Recreate a Culture of Organizing," *OCAW Reporter* 49, nos. 1–2 (March/April), 5.

Oil, Chemical and Atomic Workers International Union Local 4-620. 1988a. *Update on the 43-Month BASF Lockout at Geismar, Louisiana.* Baton Rouge: Oil, Chemical ant Atomic Workers International Union Local 4-620 [January 14].

Oil, Chemical and Atomic Workers International Union Local 4-620. 1988b. *Basagram: News and Views of OCAW Local 4-620.* Volume 12 [January].

Oil, Chemical and Atomic Workers International Union Local 4-620. 1989. *OCAW Observer: News and Views of OCAW Local 4-620.* Baton Rouge: Oil, Chemical and Atomic Workers International Union Local 4-620 [April 18].

Oil, Chemical and Atomic Workers International Union Local 4-620. 1991. *The Observer: A Union Newsletter of the Oil, Chemical and Atomic Workers International Union Local 4-620 Representing Chemical Workers at BASF Chemicals and Vulcan Chemicals, Geismar, Louisiana.*

Oliver, Pamela E. 1989. "Bringing the Crowd Back In: The Non-Organizational Elements of Social Movements." In Louis Kriesberg (ed.) *Research in Social Movements, Conflicts and Change: A Research Annual.* Greenwich: JAI Press, 1–30.

Padget, Mike. 1970. "Rites on Earth Day Join Many Together." *Albuquerque Journal* 364, no. 23 (April 23), A1–A2.

Paehlke, Robert and Pauline Vaillancourt Rosenau. 1993. "Environment/Equity: Tensions in North American Politics." *Policy Studies Journal* 21, no. 4 (Winter), 672–686.

Paterson, Kent. 1991. "Will Green Teams Widen Their Spectrum?" *Voces Unidas* 1, no. 2 (Spring), 1.

Pena, Devon. 1992. "The 'Brown' and the 'Green': Chicanos and Environmental Politics in the Upper Rio Grande." *Capitalism, Nature, Socialism* 3, no. 1 (March), 79–103.

Pulido, Laura. 1994. "Restructuring and the Contraction and Expansion of Environmental Rights in the United States." *Environment and Planning* 26, no. 5 (May), 915–936.

Pulido, Laura. 1996. "Ecological Legitimacy and Cultural Essentialism, Hispano Grazing in The Southwest." *Capitalism, Nature, Socialism* 7, no. 4 (December), 37–58.

Ramsey, Kikanza. 1993. "Year After Blast, Watchdog Fights Texaco." [Wilmington] *News Pilot* (September 15).

Ross, Andrew. 1994. *The Chicago Gangster Theory of Life: Nature's Debt to Society*. London: Verso.

Ruffins, Paul. 1989a. "Blacks Suffer Health Hazards Yet Remain Inactive on Environment." *The Los Angeles Times* (August 27), E3.

Ruffins, Paul. 1989b. "Black Americans: Awakening to Ecology." *The Los Angeles Times* (December 24), C1.

Ruffins, Paul. 1991. "Environmental Commitment as if People Didn't Matter." In Bunyan Bryant and Paul Mohai (eds.) *Environmental Racism: Issues and Dilemmas, a University of Michigan Symposium on Martin Luther King*. Ann Arbor: University of Michigan Office of Minority Affairs, 51–57.

Russell, Dick. 1989. "Environmental Racism: Minority Communities and Their Battle Against Toxics." *The Amicus Journal* 11, no. 2 (Spring), 22–32.

Sagoff, Mark. 1990. "I Am No Greenpeacer But . . . or Environmentalism, Risk Communication and the Lower Middle Class." In W. Michael Hoffman, Robert Frederick and Edward S. Petty (eds.) *Business, Ethics and the Environment: The Public Policy Debate*. New York: Quorum Books, 101–123.

Schleifstein, Mark. 1988. "March to Take Steps Against Pollution." *The Times-Picayune*. (November 1), B5.

Schleifstein, Mark. 1993. "Environmentalism Racism Targeted, Summit with Clinton Sought." *The Times Picayune* (April 8).

Schnaiberg, Allen. 1980. *The Environment: From Surplus to Scarcity*. New York: Oxford University Press.

Schneider, Keith. 1993. "The Regulatory Thickets of Environmental Racism." *The New York Times* (December 19), D5.

Schwab, Jim. 1994. *Deeper Shades of Green: The Rise of Blue-Collar and Minority Environmentalism*. San Francisco: Sierra Club Books.

Shabecoff, Philip. 1990. "Environmental Groups Told They Are Racist in Hiring." *The New York Times* (February 1), A20.

Shabecoff, Philip. 1993. *A Fierce Green Fire: The American Environmental Movement*. New York: Hill and Wang.

Sierra Club Delta Chapter. 1986. *A Preliminary Report to the Louisiana House Natural Resources Subcommittee on Oversight Concerning the Annual Dumping of Almost 200 Million Pounds of Toxic Chemicals into the Air in the Geismar and St. Gabriel Communities by 18 Petrochemical Plants and Numerous Accidental Toxic Chemical Releases During the First Nine Months of 1986*. New Orleans: Sierra Club Delta Chapter [October 21].

Smith, Damu. 1995. "Justice Time." *Crossroads* 51 (May), 15–17.

Snow, David A. 1979. "A Dramaturgical Analysis of Movement Accommodation: Building Idiosyncrasy Credit as a Movement Mobilization Strategy." *Symbolic Interaction* 2, no. 1 (Spring), 23–44.

Snow, David A. and Robert D. Benford. 1988. "Ideology, Frame Resonance and Participant Mobilization." In Bert Klandermans, Hanspeter Kriesi and Sidney Tarrow (eds.) *From Structure to Action: Comparing Social Movement Research Across Cultures*. Greenwich: JAI Press, 197–217.

Snow, David A. and Robert D. Benford. 1992. "Master Frames and Cycles of Protest." In Aldon D. Morris and Carol McClurg Mueller (eds.) *Frontiers in Social Movement Theory*. New Haven: Yale University Press, 133–155.

Snow, David A. and Pamela E. Oliver. 1995. "Social Movements and Collective Behavior: Social Psychological Dimensions and Considerations." In Karen S. Cook, Gary Alan Fine and James S. House (eds.) *Sociological Perspectives on Social Psychology*. Boston: Allyn and Bacon.

Snow, David A., E. Burke Rochford, Jr., Steven K. Worden and Robert D. Benford. 1986. "Frame Alignment Processes, Micromobilization and Movement Participation." *American Sociological Review* 51, no. 4 (August), 464–481.

Snow, David A., Louis A. Zurcher and Sheldon Ekland-Olson. 1980. "Social Networks and Social Movements: A Microstructural Approach to Differential Recruitment." *American Sociological Review* 45, no. 5 (October), 787–801.

Southern Organizing Committee for Economic and Social Justice. 1992. *Southern Community/Labor Conference for Environmental Justice Bulletin and Update, September 1992*. Atlanta, Georgia: Southern Organizing Committee for Economic and Social Justice.

Southern Organizing Committee for Economic and Social Justice. 1993a. *In Communities Across the South . . .* Birmingham: Southern Organizing Committee for Economic and Social Justice [Letter, Fall 1993].

Southern Organizing Committee for Economic and Social Justice. 1993b. *The Environmental Justice Movement in the South: Its Scope, Its History, Its Characteristics and Needs*. Atlanta, Georgia: Southern Organizing Committee for Economic and Social Justice.

Southwest Network for Environmental and Economic Justice. 1991. *Southwest Network for Environmental and Economic Justice: A Multiracial, Multicultural Organization of Community Based Grassroots Organizations and Activists from Throughout the Southwest Confronting Issues of Environmental and Economic Racism*. Albuquerque, New Mexico: Southwest Network for Environmental and Economic Justice.

SouthWest Organizing Project. 1990a. *Letter to the National Wildlife Federation, Sierra Club, National Audubon Society, Environmental Defense Fund, Environmental Policy Institute, Friends of the Earth, Izaak Walton League,*

National Parks and Conservation Association, and Natural Resources Defense Fund. Albuquerque, New Mexico: SouthWest Organizing Project [March 16].

SouthWest Organizing Project. 1990b. *Major National Environmental Organizations and Problems of the "Environmental" Movement.* Albuquerque: SouthWest Organizing Project [February].

SouthWest Organizing Project. 1991. *1981–1991: A Decade of Struggle for Social Justice.* Albuquerque, New Mexico: SouthWest Organizing Project.

SouthWest Organizing Project. 1993a. *Report on the Interfaith Hearings on Toxic Poisoning in Communities of Color.* Report delivered at Menaul High School, Albuquerque, New Mexico [April 3].

SouthWest Organizing Project. 1993b. *Campaign Work Review, SouthWest Organizing Project 1993.* Albuquerque, New Mexico: SouthWest Organizing Project.

SouthWest Organizing Project. 1994. *Development, Environment and Race in New Mexico.* Albuquerque, New Mexico: SouthWest Organizing Project.

SouthWest Organizing Project. 1995. *Intel Inside New Mexico: A Case Study of Environmental and Economic Injustice.* Albuquerque, New Mexico: SouthWest Organizing Project.

Stoecker, Randy. 1992. "Community, Movement, Organization: The Problem of Identity Convergence in Collective Action." *The Sociological Quarterly* 36, no. 1 (Winter), 111–130.

Stone, Pamela K. and Lois Marie Gibbs. 1992. "Corporate Tax Breaks: The Real Welfare Scam." *Everyone's Backyard* 10, no. 6 (December), 3.

Strasser, Fred, Anthony Paonita, Joseph Phalon and Mare Desmond. 1992. (eds.) "Unequal Protection: The Racial Divide in Environmental Law." *The National Law Journal* 15, no. 3 (September 21). Supplement.

Suro, Roberto. 1993. "Pollution-Weary Minorities Try Civil Rights Tack." *The New York Times* (January 11), A1.

Szasz, Andrew. 1994. *EcoPopulism: Toxic Waste and the Movement for Environmental Justice.* Minneapolis: University of Minneapolis Press.

Tarrow, Sidney. 1994. *Power in Movement: Social Movements, Collective Action and Politics.* Cambridge: Cambridge University Press.

Taylor, Dorceta E. 1989. "Blacks and the Environment: Toward an Explanation of the Concern and Action Gap Between Blacks and Whites." *Environment and Behavior* 21, no. 2 (March), 175–205.

Truax, Hawley. 1990. "Beyond White Environmentalism: Minorities and the Environment." *Environmental Action* 21, no. 4 (January/February), 19–30.

United Church of Christ Commission for Racial Justice. 1987. *Toxic Wastes and Race in the United States: A National Report on the Racial and Socio-*

economic Characteristics of Communities with Hazardous Waste Sites. New York: United Church of Christ.

United Church of Christ Commission for Racial Justice. 1992. *Proceedings of the First National People of Color Environmental Leadership Summit.* New York: United Church of Christ Commission for Racial Justice.

U. S. General Accounting Office. 1983. *Siting of Hazardous Waste Landfills and Their Correlation With Racial and Economic Status of Surrounding Communities.* Washington, D.C.: Government Printing Office.

Van Liere, Kent D. and Riley E. Dunlap. 1980. "The Social Bases of Environmental Concern: A Review of Hypotheses, Explanations and Empirical Evidence." *Public Opinion Quarterly* 44, no. 2 (Summer), 181–197.

Williams, Raymond. 1984. *Problems in Materialism and Culture.* London: Verso.

Wilson, Rand. 1990. "An Innovative Joint Campaign." *LRA's Economic Notes* 58, no. 9–10 (September/October), 10–11.

Wright, Beverly H., Pat Bryant and Robert D. Bullard. 1993. "Coping with Poisons in Cancer Alley." In Robert D. Bullard (ed.) *Unequal Protection: Environmental Justice and Communities of Color.* San Francisco: Sierra Club Books, 110–129.

Zeff, Robin Lee, Marsha Love and Karen Stults. 1989. (eds.) *Empowering Ourselves: Women and Toxics Organizing.* Arlington, Virginia: Citizen's Clearinghouse for Hazardous Wastes.

Zwerdling, Daniel. 1973. "Poverty and Pollution." *The Progressive* 37, no. 1 (January), 25–29.

Index

About the Author

PATRICK NOVOTNY is an Assistant Professor in the Department of Political Science at Georgia Southern University. He has published in journals including *The Harvard International Journal of Press/Politics, Social Science Computer Review; New Political Science; Peace and Change; Social Justice;* and *Capitalism, Nature, Socialism.* He is currently working on a study of the history of education and civil rights in the state of Georgia during the early 1940s.